The turner's co[m]

containing instructions in concentric, [e]llip[tic], [a]nd eccentric turning; also various plates of chucks, tools and instruments: and directions for using the eccentric cutter, drill, vertical cutter, and circular rest; with patterns, and instructions for working them.

Anonymous

Alpha Editions

This edition published in 2024

ISBN : 9789362515063

Design and Setting By
Alpha Editions
www.alphaedis.com
Email - info@alphaedis.com

PREFACE.

The primary object of the author, in offering THE TURNER'S COMPANION to the notice of the public, is the hope of explaining, in a clear, concise, and intelligible manner, the rudiments of this beautiful art; an art immortalized by the pen of Virgil, practised by the Greeks and Romans, and, as we are told, still existing in those exquisite *chef-d'œuvres* of former years, so much admired and sought after in our century.

A short treatise on this subject is much wanted, and the author has endeavoured to give such correct and comprehensive information as will, he hopes, render the following pages useful as a book of practical instruction to the beginner, and of reference to those already advanced in the study of this beautiful science. The activity of mind requisite for the attainment of perfect success in all the various branches of Turning, by exercising the inventive and reflective powers, cannot fail of producing a beneficial effect on the character of youth, which must prove a lasting advantage; and the minute accuracy necessary for handling the tools, serves to confirm a steadiness of sight and hand that must, in after years, and in other branches of science, be highly appreciated.

From a long experience in the endless sources of interest and occupation derivable from this pleasing and salutary employment, the author feels assured that whoever has once patience and perseverance to overcome the first difficulties, will speedily, like himself, become an enthusiast in the art. And why should not our fair countrywomen participate in this amusement? Do they fear it is too masculine and laborious for a female hand? If so, that anxiety is easily removed; the rough work can be executed by any carpenter, and when once prepared, what occupation can be more interesting and elegant than ornamenting wood or ivory in delicate and intricate patterns, and imitating, with the aid of the lathe, the beautiful Chinese carving, so much and so justly admired? Besides, the taper fingers of the fair sex are far better suited than a man's heavier hand, to produce the requisite lightness and clearness of effect. To our charitable countrywomen, who employ so much of their time in raising funds for the diffusion of Christianity in far distant lands, and for augmenting the comforts of the poor in our own happy land, the lathe will prove a most useful auxiliary, as well as to those who are anxious to bestow beautiful and cherished remembrances on absent friends.

Another very forcible argument in favour of the amusement of Turning being cultivated by the ladies and gentlemen of our free and independent country, and one which will, I am sure, plead most strongly with all parents and guardians, is, that all occupations within doors being usually of a sedentary nature, the exercise attendant upon the use of the lathe must prove highly beneficial to health; and one moment's reflection will point out the incalculable advantage to be derived from instilling the love of useful employment in every youthful mind; or, when freed from the irksomeness of graver study, may they not seek companions and pursuits to whom they would ever have remained strangers, had their idle hours not hung heavy on their hands?

Should this little work prevail upon any of my readers to commence the study of this truly beautiful science, and should the occupation, as no doubt it must, prove a source of pleasure by adding to their amusement, it will amply repay the labour it has required, and bestow sincere gratification on the Author.

THE LATHE.

───────

"The pride of arts from fair Ambition springs,

And blooms secure beneath her fostering wings,"

───────

Among all the many descriptions of the varied, beautiful, and useful inventions that owe their discovery and perfection to the genius and hand of man, inventions that in so wonderful a degree assist and facilitate the operations of the mechanic, no one has ever written the history of the Lathe. It seems strange that in a land where mechanism is carried to its greatest extent, where science of every kind is fostered and encouraged, the beautiful machinery, the easy management, and wonderful precision obtained by the aid of the Lathe, have never yet, in our language, found a pen willing to describe them. And yet, to the architect, the mathematician, the astronomer, and the natural philosopher, Turning is as useful, nay indispensable, as to the watchmaker, the goldsmith, the joiner, and smith. And it is not by these alone that its powers are appreciated; many of those who by birth, station, and riches are not in a situation to require its aid in their scientific and mechanical operations, still find this art, from its great simplicity, from the perfect ease and accuracy with which the most delicate and intricate workmanship is performed, the agreeable occupation it gives to the mind, and the beauty, elegance, and utility of its products, one of the most interesting and healthful that can be followed.

The treasures of all lands are converted into various and beautiful articles by the aid of the lathe. Gold and silver, brass, iron, and copper,—the magnificent trees that grow in the deep forests of the West, and those that flourish on the burning plains of Africa—the ivory obtained from the tusks of the elephant and hippopotamus,—the coal, jet, alabaster, and marble, dug from the bowels of the earth—are all of the greatest value to the turner. A kind of cocoa nut has also lately been brought from the West Indies, which, being hard, white, and tough, renders it excellent for working in the lathe; when polished, it has the appearance of a substance between ivory and mother-of-pearl.

To the Greeks and Romans (for the exact place of its origin is not known) the invention of this ingenious machine is ascribed; and though, doubtless, in our time it has been greatly improved and perfected, still the ancients, to whom we owe so much, first discovered and used it; and by them its powers were so well appreciated, that we are told it became a proverb among them to say any thing was formed in the lathe to express its justness and accuracy.

The Greek and Latin authors frequently mention it in their writings, but they have not clearly handed down to posterity the name of the first inventor; indeed, on this point there are many and varied opinions. The Sicilian historian, Diodorus Siculus, informs us that the first person who made use of the lathe was a nephew of Dædalus, by some authors named Talus, by others Perdix. This youth, we are told, invented the saw, compasses, and other mechanical instruments; and to him we possibly are indebted for the lathe also, for we are told, in ancient mythological history, so great were his ingenuity and talent for invention, that his genius soon surpassed even that of his uncle, who, enraged at his celebrity, and jealous of his rising fame, scrupled not to sacrifice him to his feelings of rage and hatred: some say he was poisoned; others, that he was precipitated from a high tower in the citadel of Athens; and the same authors assure us he was changed into a partridge.

Pliny, however, (and his words are great authority), ascribes the invention of the lathe to Theodore, of Samos, an artist who discovered the method of melting iron, of which he made statues. The same author also mentions a man of the name of Thericles, who was celebrated for his dexterity in Turning; and Virgil says—

"Lenta quibus torno facili superaddita vitis."

These testimonies, of ancient poets and historians conjoined, prove it to be an art of the greatest antiquity. Cicero also mentions it; and it is affirmed that, with this machine, the Greeks and Romans turned all kinds of urns and vases, and adorned them with ornaments in basso relievo. If to Turning we really owe those treasures of other days which are found buried among the ruins of Herculaneum and Pompeii—treasures so valued by the antiquary as a memorial of former ages; by the lover of the fine arts as beautiful and graceful additions to his cabinet of curiosities; and by the rich and opulent as ornaments superior to any that modern hands can produce;—surely we must confess the workmanship of the lathes of our times is not to be compared with that performed by the more simple machinery of centuries ago. It seems, indeed, almost miraculous, that the beautiful figures and elegant and graceful designs here spoken of should be produced by a potter's wheel, so was the lathe anciently denominated; but the testimonies of so many learned historians agree in declaring that to its aid we owe those exquisite productions, that it is impossible for even the most skeptical to deny it.

Before we quit, what may not unaptly be denominated, the romance of the history of the lathe, we will add, that the saw, which we have already said was invented by Talus, is supposed to have been first made by him in imitation of either the jaw-bone of a snake, or else the back-bone of a fish; and in a painting still preserved among the antiquities of Herculaneum, is a saw exactly resembling our frame saw, with which two genii are dividing a piece of wood.

It is, however, certain, that could these Roman and Grecian artificers see a modern lathe, examine its complex yet beautiful mechanism, and the almost endless additions and improvements it has undergone since the days of Virgil and Pliny, they would hardly recognise it in its more finished state, and would be much puzzled to discover in what manner to manage its machinery. Among the numerous apparatus adjusted to it is a machine, by the aid of which medallions have been executed; and in the British Museum is a profile in basso relievo of Sir Isaac Newton, wholly worked in the lathe; but how different from the turning of the ancients! The medallion machine requires much labour and very expensive apparatus, while the potter's wheel cannot have possessed much mechanism or great quantities of tools: most of those now in use being unknown in former days.

Having now established the great antiquity of the lathe as a useful and classical employment, we will briefly mention a few of the improvements it has undergone in latter years. The potter's wheel is, of all lathes, the most simple; it merely consists of an iron beam, or axis, a small wooden wheel placed on the beam, and a larger one fastened to the end of the same beam, which turns by a pivot on an iron stand. With this simple contrivance the workman still forms the body of the vessel of clay, but never attempts to turn the handles, feet, mouldings or ornaments. In latter years, various and important tools and improvements have been made; chucks have been invented, which enable the turner to accomplish with speed and facility an almost innumerable variety of circles, lines, ellipses, and arcs, all so delicate and true in their form and design, that they cannot fail exciting the admiration and wonder of all who contemplate them. The screw also, once a formidable difficulty to the uninitiated, is now rendered perfectly easy of execution by means of the traversing mandrel; those, too, who are very learned in this art, can out of a piece of ivory or mother-of-pearl, produce in the lathe beautiful brooches, ear-rings, and studs, worked in raised flowers; chessmen in imitation of carving, and ornamented vases full of detached flowers; while fluted and spiral columns, delicate mouldings, and fanciful beadings, are of comparatively easy execution.

To form patterns upon wood or ivory, various descriptions of chucks are employed; one lathe serves for all, as they are made to screw on to the nose of the mandrel. By the aid of the concentric or common chuck, every article you turn is circular; the lines forming the circle are enlarged or decreased as the tool approaches or recedes from the axis. The oval chuck, as its name signifies, works designs of an oval or spherical shape; the eccentric turns patterns of a circular form, but its peculiar properties enable the workman to alter the centre of his work at pleasure: the geometric and compound eccentric produce beautiful geometric and carved designs; the oblique and the epicycloidal also turn curious and intricate patterns, and the straight line chuck performs all its work in direct lines. These are the chucks most in use, but many of them are expensive and complicated, and they only execute the ornamental work; the shape and size of the object are accomplished by the lathe, without any aid but that of a common chuck and common tools.

The curious and varied mechanism of the above-mentioned chucks are truly wonderful, and the patterns they perform very beautiful; one of peculiar form, and exceedingly intricate, has been invented for bankers' checks, to prevent forgery. There are also two rests, which are necessary appendages to them; the sliding rest, that moves in a direct line at any angle, and the circular rest, which enables the turner to ornament balls, spheres, and round objects. We must not omit, too, to mention the eccentric cutter, the drill, and the universal cutter, all exceedingly useful, and enabling the turner to execute a great variety of designs and patterns. The rose engine, also, is much admired for all kinds of ornamental work, but it is very expensive, and new inventions are daily adding to the machinery of the lathe, and rendering its powers more extensive.

The wonderful discovery of voltaic electricity, by which copper-plates, plaster casts, wood engravings, and medals may be copied, can also be applied in various ways to turning, either in wood or ivory. For instance, by its aid a wooden thimble may be changed into the resemblance of gold, or a box take the appearance of silver; thus, while your work retains its first beauty and delicacy, the material is apparently of much value. It would be going beyond the bounds of a work on turning, to give any directions for this transmutation, particularly as there are so many already published on the subject. We must also add, that by a careful and steady management, the drill may be made almost to take the place of the graver; by holding a plate of copper steadily against it, and using various tools, (not letting the lathe go too quick,) portraits and landscapes can be executed for printing.

Having now finished this short sketch, which I hope will not be unacceptable to our readers, and will perhaps induce them to follow this interesting and healthful occupation, we must beg that those who peruse "The Turner's Companion," and follow the directions it contains, will not be daunted by the first difficulties that assail them, but will patiently persevere till experience enables them to overcome and vanquish them, remembering that—

"The wise and active conquer difficulties,

By daring to attempt them. Sloth and folly

Shiver and shrink at sight of toil and hazard,

And make th' impossibility they fear."

LIST OF TOOLS NECESSARY FOR TURNING.

Gouges.	Screw-driver.
Chisels.	Pincers.
Scrapers.	Compasses.
Side tools.	Rule.
Point tools.	Callipers.
Moulding tools.	T square.
Inside tools.	Brace and Bits.
Planes.	Screw tools.
Drills.	Milling tools.
Hatchet.	Oil-can.
Mallet.	Glue-pot.
Hammer.	Sand Paper.
Files.	Chalk.
Vice.	Glue.
Hand Vice.	Isinglass.
Gimlets.	Pumice Stone.
Saws.	Nails.

THE TURNER'S COMPANION.

The machines used for Turning, whether round or oval objects, are called Lathes; they are of various shapes and sizes; some very small, as those generally used by watchmakers; others very large and powerful, for turning iron; and others, the kind I am going to describe, of a middling size, for fashioning wood and ivory. The large lathes, being too heavy to be worked with the foot, are usually turned by a steam-engine, but the foot-lathe is the most convenient for the turner in wood; it may be made of iron or wood: if of the latter, it should be constructed entirely of very hard, well-seasoned oak, or of mahogany. There are various opinions respecting the advantages and disadvantages of metallic and wooden lathes; in the former, it is impossible to obviate an elastic tremor, which is unpleasant and injurious; but then, on the other hand, they are so much more durable and compact; and they enable you to perform your work with so much more accuracy and exactitude, that they are, on the whole, perhaps, to be preferred. The drawing given in Plate 1 will serve as a pattern for either an iron or a wooden lathe; but as the workman could construct the latter for himself, we will suppose the description we are about to give relates to a wooden one.

The bed of the lathe, B B, may be of any length required, and is firmly fastened with bolts to the uprights O O, which form the legs of the lathe, and to which the bed is strongly attached by bolts passing through both; while the nuts that draw them tight, being what is called *countersunk*, are of no inconvenience to the workman. The feet and the two uprights must also be firmly fastened to the legs O O; and to prevent the least unsteadiness or motion, they must be screwed strongly to the floor, and must be of a sufficient size to form a solid support to the lathe.

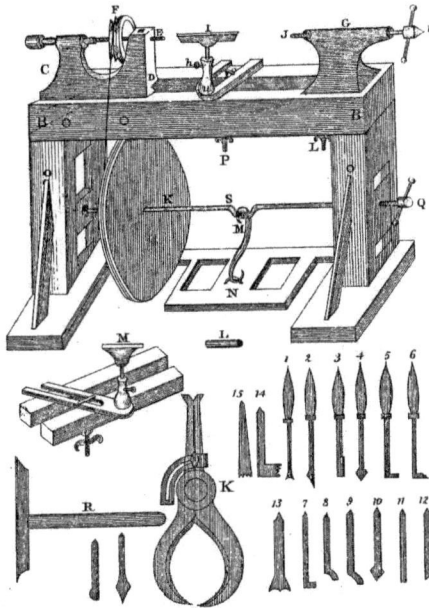

Plate 1.

The left hand puppets, C D, or, as they are sometimes called, the headstock, should be of iron, and cast in one piece. The under part fits tightly into the open space in the bed of the lathe, and is fixed there with screws; while the two cheeks of the puppets rest *on* the bed itself. The mandrel E, to prevent, as much as you can, any vibration, should be as long as possible; it runs in a metal collar, through the puppets C D, and is of steel, turned perfectly cylindrical; it is kept constantly oiled, by pouring a few drops of oil upon it through holes made in each of the puppets. The screw E, at the end, is called the nose of the spindle, and upon it the chucks intended to receive the work are screwed. The back puppet, G, is used to support long pieces of wood; it is moved backwards and forwards on the bed of the lathe, so as to suit the work upon which you are occupied, by loosening the screw L; within the upper part of G is a steel spindle, J, which screws in and out of the headstock, by turning the screw K. Care must be taken that the point of this spindle be exactly on a line with the nose of the mandrel, E. The point, J, takes out, and another nose, L, can be inserted in its place, to receive the pointed end of any small work, should it be more convenient.

Upon the spindle is a brass or a mahogany wheel, F; it has three grooves in it, and the great wheel, K, has three similar grooves turned in a V, so as more effectually to take hold of the band which moves them round. The three different grooves in this wheel, and in the small one, give different velocities to them. The band which turns them is made of strong catgut, and passes under the lower and over the upper wheel, working in the corresponding grooves of each; it is joined with a hook and eye of iron, that have a screw in them. Slightly taper off with a sharp penknife a little of each end of the catgut, so that it will just enter the hook and eye; then hold the band firmly in a vice with your left hand, and with your right take up the hook or eye in a pair of pincers, and screw it upon the catgut till quite firm. This is a far better means of joining the band than any other that can be employed, as the hooks and eyes seldom give way, and obviate the necessity of knots or joins, which are always clumsy and inconvenient. Cord, too, gives way with the variations of the atmosphere, so that it constantly requires shortening in dry, and lengthening in rainy weather; catgut is so slightly influenced by these changes, that its use is far preferable. The treadle, N, when moved up and down with the foot, gives motion to the two wheels, and thus the spindle, with the wood to be worked screwed upon it, is turned round with a quick or slow movement. The axle of the great wheel, S, works in two screws, Q; the crank, M, is connected at one end with the axle, S, and at the other hooks into the treadle frame, N. Plenty of oil should be given to the axle at Q to enable it to work easily. The rest which supports the tools is represented at Fig. 1, and M; it is made of iron, and consists of three parts; the lower has a forked foot, T, which rests upon the bed of the lathe, and enables it to be drawn backwards and forwards, so as to accommodate the workman; this foot is held in its place by a bolt, O, which, passing through the bed of the lathe, is sufficiently broad to rest upon each side of the foot, T, and is tightened by a screw that passes underneath the lathe, P, through which it passes; the upper part of the rest is a cross piece of iron with a cylindrical stem, that fits into the socket, U, and is moved up and down, to the right or the left, by loosening the screw, H. The workman should have rests of different sizes, to suit various kinds of work, but they must all fit into the same socket.

Having now given a clear description of a lathe, I need only add, that it should be placed opposite a window, so as to have the benefit of as much light as possible; a skylight above the head is also a great advantage. As it is indispensably necessary for the learner to exercise himself in plain turning, that is, in the formation of different articles, so as to be able to turn them perfectly round, oval, or hollow, as required, before attempting more difficult and complicated work, we will now give a list of the most useful tools for this purpose.

For the wood, the gouge, Fig. 1, plate 1, is first to be employed, to reduce the unevenness of your work; its edge is rounded. To use it, place the rest on a level with the axis of the work, and hold the handle of the tool downwards so that its cutting edge is *above* the axis. These tools are useful for making concave mouldings. In using them, do not push them roughly against the wood till it becomes tolerably even, or you will spoil their edge and chip the work; and hold your hands very steady.

The chisel, Fig. 2, is next used, to give a smooth and polished appearance to the wood. Its cutting edge is oblique. Elevate the rest considerably *above* the axis of the work, so that, though held with a less inclination than the gouge, the edge of the chisel operates on a higher part of the surface. Use this tool at first with great caution, for it is much more difficult to manage than the gouge; with an inexperienced hand, the point is apt to dig into the wood, quite spoiling its surface, or else, by pressing it too firmly upon the work, it cuts great pieces in an uneven manner. When skillfully used, it should feel almost to work by itself, merely running steadily along the wood, shaving off all its inaccuracies, and making it look quite bright, smooth, and polished. All soft woods are entirely turned with the gouge and chisel, of both of which you must have several sizes.

Fig. 3 is called a right-side tool, and has two cutting edges, a side edge and an end edge: so as at the same time to cut the bottom and side of a cavity. The left-side tool cuts with the opposite side. In using them, hold the bevel which forms the edge downwards.

Fig. 4 is a point tool, useful for making small mouldings, and much employed in finishing the shoulders and flat ends of work.

Figs. 5, 6, 7, 8, 9, are inside tools, used to turn out hollows; also to make cups, and various other articles.

Fig. 10 is a parting tool, used to cut off work, and to make incisions.

Fig. 11 is used exclusively for very hard woods, as cocoas and ebony, which chip if attempted to be smoothed with a chisel; also for turning ivory, bone, or jet; one side and the end are sharp. This tool is very strong, and requires some practice to use it well. Be very careful, in sharpening it, to keep the front edge quite straight, or else, in hollowing out boxes, the inner sides will not be turned out evenly; that is, one part will be thinner than another. This tool is held flat upon the rest, which must be on a level with the axis of the work, or sometimes the tool, by raising the handle, may be lowered so as just to scrape the wood. It is frequently called a graver; in turning metals, it is the tool first used. Copper and brass are easy to turn, and in case of necessity, it is useful to have the proper tools, and to know how to manage them.

These tools all are indispensably necessary to the turner, and he should exercise himself constantly in their use; for until he becomes quite master of them, he will injure, spoil, chip, and destroy, whatever he attempts to turn.

Figs. 12 and 13 are very useful to make mouldings of various kinds.

The handles of the tools must be made of very hard wood, and it is a great convenience to have them all of nearly the same size; for the hand, getting accustomed to them, manages them with more facility. Drive the tool firmly into the handle, and hold it there by a broad brass ring, as in No. 1.

To keep the tools in good order, that is, properly ground and sharp, demands great attention. If they become chipped, grind them even on a grindstone, taking care that the BEVELS retain their proper angles. To avoid spoiling their edges, and to enable you to have them always near at hand, a rack perforated with holes, into which they can slip, is very useful. This rack may be fastened against the wall, near the lathe. A screw-driver, two or three different-sized gimlets, and nails of various kinds, must always form part of the turner's tool-box; also some files, and a hand-vice.

Before we leave the description of turning tools, we will mention the saw, which is an indispensable addition to the tool-box. There are many kinds, but the most useful are the hand-saw, the tenon saw, and the circular saw. The first is about twenty-six inches long, and is generally made with four teeth to an inch. It is used for cutting wood across, and in the direction of its fibres. The teeth at the lower end are smaller than the upper ones, by which means the wood is not so much torn as if the teeth were all of an equal size. The tenon saw is used for cutting across the fibres of wood; the smallest saw of this kind is about fourteen inches long, the largest about twenty inches. Circular saws are of all sizes; they are easily fitted up with a spindle, which, being screwed on to the nose of the mandrel, and supported at the other end by the back puppet, enables the workman to turn them by the wheel of the lathe, while at the same time he holds the wood or ivory firmly against them. Should a larger circular saw be required, it is more advisable to fit it up separately from the lathe, with a frame-work and wheel to itself.

Glue, which is very necessary for turners, requires some little care in preparing; it must first be steeped for several hours in cold water to soften it; if it swells without melting, it is good, and must then be dissolved in water; the proper quantities are, a quart of water to half a pound of glue. The heat should be just enough to melt it, and the pan in which it is contained must be placed in a larger copper vessel, filled with water; by this means, when the water in the outer pan boils, the glue will be dissolved without any fear of its burning, which would immediately spoil it. When you are going to glue a piece of wood to a chuck, put very little glue thinly and evenly over the surface of the latter, then press the wood upon it firmly, and place a lead weight upon them to unite them perfectly.

CHUCKS.

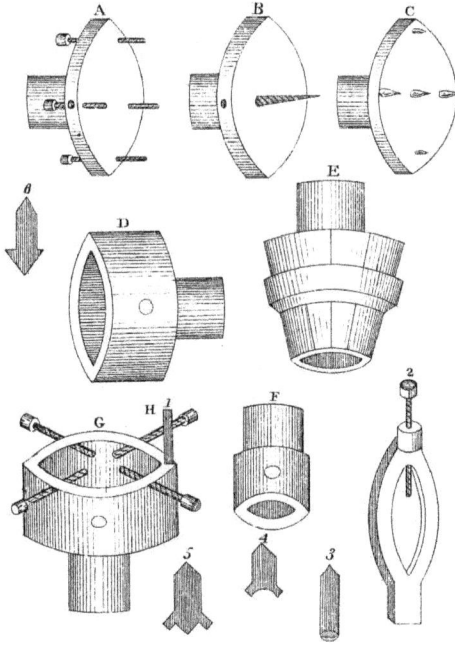

Plate 2.

The chucks, upon which the material to be turned, whether of wood, ivory, or metal, is always fixed, next demand our attention. They are of every variety of size and form, and are all screwed upon the mandrel of the lathe. Many are made of brass; others (the most numerous, because the turner can make them for himself), are of wood; but these latter should be used soon after they are made; for if not constructed of very dry, hard wood, any great variation in the weather will cause them to shrink, and thus the screw becomes slightly altered, and will not fit tightly to the neck of the mandrel. To avoid the expense of having many brass chucks, which would be very great, if we procured them of the size and shape requisite for all kinds of work, it is a good plan to have several brass plates made about the size of half-a-crown, plate 2, A, with a screw in them, to fit upon the mandrel, and four screw-nails with which to fasten them to wooden chucks of any form. Thus, when these chucks are worn out, unscrew the brass plates and screw them on to others. In plate 2 are the drawings of several chucks, which are useful for various purposes; they are all made of brass, with a screw that fits upon the mandrel. B is a brass plate, about two inches in diameter; from the middle projects a tapering screw, about half an inch long. This is used to hold any thing that is flat, as a stand, or candlestick base, or, with the aid of the back puppet, to support a long piece of wood, while turning down to fit a stronger chuck. For this purpose, bore a hole in the wood, and screw it on to the chuck. C is the same shape, but has five iron points projecting from its surface; upon them the wood must be firmly fixed by hammering it on. It is better to use the back puppet, as the wood is apt to become loosened by a sudden jar, or any unevenness in the surface.

D is universally useful, either for large or small pieces of work. The wood or ivory may be turned to the proper size to fit, on the chuck B, and then driven firmly into the hollow cup with a wooden mallet; or a piece of common wood may be made to fit it tightly, and a hole turned in it to hold the object you intend to turn.

E is called a ring chuck, and is made of box wood. Drill a hole through the centre, and then saw it across in six parts. By its being turned smaller at one end than at the other, this chuck opens at the *sawgates*; you then drive on to it, with a hammer, a ring of metal, and the wood inserted in the hole will remain immovable.

F is called a square hole chuck, the hole in the middle having several drills and bits to use with it, as in 3, 4, 5, 6. G is called a die chuck: it is the same in shape as the cup chuck, only not so deep, and it has several screws passing through its sides at equal distances, and meeting in the centre, by which the work is held, so that it serves equally for a large or a small object.

H is exactly the same in shape as G; with the addition of an *arm*, No. 1, the use of which is as follows. If this chuck does not turn the wood round properly, fasten to the latter what is called a *carrier*, No. 2, the end of which, projecting further than the chuck, rests upon the arm, 1, and causes all to turn together.

These appear to us to be the chucks most universally in use for concentric turning; the ordinary ones the turner may make of wood, and those for ornamental work we shall mention hereafter.

It may, however, be as well to describe the method of making the wood chucks:—Select a piece of close-grained dry wood—box is the best; having taken off the corners and made it tolerably round with a chisel, or a small hatchet, you must then find the centre of the two ends. To do this, lay the piece of wood on a bench; open a pair of compasses to nearly half the diameter of the piece; fix one point of the compasses firmly in the middle, and with the other draw a circle as near the edge of the wood as you can. If you find the circle is not exact, but further from the edge on one side than the other, alter the position of the compasses, till they become right. Bore a hole in the centre, when found, and screw the wood on to the chuck, B. Place your rest *facing* the work, and cut in the centre of it a hole, the depth and NEARLY the size of the screw on the nose of the mandrel. This done, take the piece of wood off the brass chuck, and fasten it firmly in a vice; then screw into the hole a TAP, which has been made on purpose to fit the screw of the mandrel. This tap cuts a thread as you turn it round in the hole, so that when you unscrew it, you have only to screw the wood to the nose of the mandrel, taking care to make it fit quite close to the shoulder; it must then be turned quite round and smooth with the gouge and chisel, and the face of it also perfectly flat, which is seen by holding against it the flat part of the T square, plate 1, R. If the square touches all the face of the chuck, it is ready to receive the wood; but if you can in any part see the light between them, take the chisel and smooth it over again. The chuck being ready, cut a piece of wood, we will say for a box, round it with a chisel, take some thin glue, and fasten the wood to the chuck with it. When quite hard and dry, begin to turn the sides even, then cut down the groove to receive the lid, which should be glued on to another chuck. To hollow out the box, turn the rest to the face of the work and use the gouge. When of a sufficient depth, take the *callipers*, plate 1, K, push the small ends down the box to the bottom, stretch them out as far as they will, and set the screw, then pull them slowly out, so as to measure the size of the top of the inside of the box; if not exactly the same, turn out a little more of the bottom. Now begin to smooth the outside and hollow out the lid, taking great care to make it fit exactly the groove made in the bottom of the box; for this purpose the callipers must be set, so that the forked end stretches to the width of the circumference of the groove; you will then find that the other end will be exactly the same width, and keep trying them to the inside of the lid, till they enter it very tightly; then try the lid on the bottom, and it will fit. By thus using the callipers, you are saved much trouble in taking the work on and off the lathe, to fit the parts together.

You must now cut the lid off the chuck, either with the parting tool, or with a saw, and having it firmly placed on the bottom, smooth and finish off the sides so that they look quite even, and as if there was no separation between them. This done, move the rest to the front of the lid, and finish it up the same, taking care not to leave the slightest scratch or unevenness on the surface; then rub the whole of the work well with sand paper, making the lathe turn very rapidly, first one way, then the other, and finish by rubbing it over with a drop of olive oil on a piece of rag, and the shavings of the wood. The great beauty of turning consists in all the parts being exact, shining, well finished off, and not too thick; to attain this latter perfection, experience is necessary, and I should advise constant practice in turning box-wood and holly, till the learner is able to make his boxes fit properly, and also look neat and light. The lid being finished, lay it aside, and saw off the bottom; then turn its own chuck to a proper size to receive it, while you smooth and polish the outer part. It is requisite to be very particular in the manner of chucking work, such as boxes, thimbles, or any thing that has been hollowed out; and remember always to make a chuck to fit INTO THEM, instead of putting THEM into one. It is also better to leave a shoulder that the work can rest against,—you are then sure that it is supported evenly; if not, you may find that the slightest inclination to one side or the other will cause you to turn one side much thinner than the other; and if you are working, we will say the top of a box, it will always look crooked, and if the bottom, it will never stand steady. Should the box feel loose and fall off the chuck before it is finished, a little chalk may be rubbed upon the former, which will give it a firmer hold, and prevent the slipperiness consequent on the friction of two pieces of wood; and sometimes a thin piece of paper inserted between them is useful in the same way; for although the work should fit close and firm on the chuck, if the chuck be *too large*, so that you have to use force to make them unite, you will most probably split your work all to pieces. Great care also is requisite to get it off the chuck; insert one of the small chisels between it and the shoulder of the chuck, and move it slightly, first on one side, then on the other. Many neat ornaments may be made on boxes with the smaller chisels, such as lines and mouldings; and there are several useful and ornamental tools, called *milling tools*, (see plate 5, Nos. 1, 2, 3), which are not expensive, and give much effect to the work. They consist of small wheels, upon which the pattern is cut. Place the rest so that there is space for the wheel to turn between it and the work, push it close up to the wood, hold the handle very firmly with both hands, so that the tool cannot slip, and with a few quick turns of the wheel of the lathe, the pattern will be clearly impressed upon the wood. With these simple tools innumerable beautiful articles may be finished; and though they require neither the application nor talent that can be displayed in performing other ornamental work with the eccentric chuck, cutter, and drill, still they possess two great advantages—cheapness, and facility of management, and are easily procured.

By a little attention and ingenuity, a great variety of elegant and useful articles may be made on the lathe, with the assistance of but a limited collection of tools,—such as thimbles, boxes, cups, rings, stands, small vases, stilettos, pen-handles, pin-cushions, needle-cases, and vinaigrettes.

THE SCREW.

We now come to the most difficult operation in turning, that of cutting a screw; to perform it well and easily is a proof of the workman's skill and proficiency in the art. There are many ways of doing it; we shall therefore give the best and least expensive. The screw tools, figures 14 and 15, plate 1, must fit exactly one into the other. Fig. 15 is an outside, Fig. 14 an inside tool. As the threads may be required to be cut coarse or fine, according to the work you are engaged upon, the small grooves in the tools are made to suit. Having turned your box quite round, and hollowed it out, cut the groove upon which the lid is to fit; place the rest at a convenient distance, turn the wheel, not too fast, and move the outside screw-tool along the rest with a regular horizontal motion, and it will cut a screw, the threads of which will fill up the space between the teeth of the tool. But care must be taken to jerk the tool off when at the end of the space intended for the screw; or if it be allowed to remain stationary, cutting the wood, the threads will be destroyed, and become useless. When this is well done, turn out the lid of the box till it nearly fits the bottom, and in the same manner press the inside screw tool against the side of the cavity, draw it out horizontally as the work moves round, and if carefully managed, it will soon be made to fit upon the outside screw. As, however, to accomplish this well, and with precision, great practice is requisite, an invention, called a traversing mandrel, is frequently used, particularly by beginners: we will endeavour to give a description of its form and use.

At the end of the mandrel A, pl. 4, is a brass cylinder, I, which fits upon the end, and is kept in its place by a nut, 2, which screws firmly into it. Below this, fixed to a brass plate that rises and lowers at pleasure by turning the screw-key, B, is attached the screw-guide; (a brass plate cut into grooves of various sizes to suit the thread you wish to cut, fig. 3;) this guide moves round on a pin. To use this machine, unscrew the nut B, pull off the brass cylinder, and in its place put on the guide, C, and screw the nut in again; turn the other guide, 3, to the groove which corresponds with it, and which is usually numbered, to avoid mistakes; turn the key B till the lower guide meets the one you have just put on the mandrel, and slips easily into it. You will now find that by only allowing the fly-wheel to move *half round* and back again, the mandrel will run backwards and forwards, and thus have the exact motion requisite for cutting a screw. Fix your rest, and hold your tool (which must have the same sized thread as the guide then on the mandrel) quite steady upon the rest, against the revolving wood, and in a few minutes the screw will be produced. The inside screw is made in the same manner, with the inside tool, by turning the rest in front of the work. Do not press the tool too hard to the wood at first, till the threads are slightly cut, so that the teeth may enter always in the same place. About six different-sized screw guides and tools to fit will be quite sufficient for an amateur turner. There is also another way of cutting screws, by means of a traversing chuck. On the mandrel, R, pl. 4, is screwed the chuck B, to which are screwed the chucks of the lathe, R. On the outside of B is turned a screw, fitted to an inside screw worked in the block C, from which extends an arm, D, sufficiently long to allow the arm E to slide up and down it; a piece of iron should be screwed to the circular block, C, of such a length as to be capable of moving in a groove that may be cut in the collar; it is intended to prevent the block C from turning quite round. The rest, G, must not stand, as usual, parallel to the work in cutting the outside screw, but at right angles, as when an inside screw is to be cut, in order that the further arm of the rest, F, may be joined to the end of the second arm, G. It is necessary that the second arm, E, shall be capable of fastening firmly the first arm, D, to any part of the rest, G, F, as also to have a joint at each end, to admit in a horizontal plane its free action. Thus, as the lathe turns to or from us, the arms must traverse forwards or backwards, which gives a similar motion to the tool, H, that is screwed firm to the further arm, F, of the rest, and thus you can cut a screw with a single point tool. Of course any unsteadiness would spoil the screw. If you draw the centre of the rest nearer to you, and thus bring the tool nearer to the arm, E, a screw of a much larger-sized thread will be cut; for as the rest, turning in its socket, moves on a centre, the further the tool is from the centre, the greater will be the radius of the circle described, and *vice versâ*.

It may, perhaps, be feared that a piece of wood so far from the collar, K, may be apt to spring; but this is easily avoided, by not making use of the chuck, B, till the screw is to be turned. Another disadvantage would seem to arise from the impossibility of cutting screws when the puppet head is made use of. But this may also be obviated by lengthening the arm, E, to the part where the screw is to be cut, and thus we have the same screw as the traversing one. The socket, S, slides on the rest, and may be fastened to it by a screw, the upper part that turns on a pivot admits the arm, E, to slide through it, which arm is held firm with a screw.

At the commencement of the work, the rest stands at right angles with the wood on which the screw must be cut; then, by bringing it back to its original angle, and sliding forwards the tool to the last thread of the screw that was just cut, we proceed to any length required. When two or three threads are cut, the most unskillful turner will be able to continue the screw with a common screw tool.

BORING COLLAR.

Fig. D, plate 4, is a boring collar, used to support any long slender body which is required to be turned hollow. Without a support of this kind it would be impossible to keep the wood in its place; and it would either incline from the centre, thus causing the hollow to be drilled out quite crooked, or it would spring from the chuck. To obviate these two inconveniences, the two collars, figs. C and D, plate 4, are employed. In the former one, which is made of iron or brass, the holes are conical, and their centres are all precisely at the same distance from the axis of the collar. In using it, remove the right hand puppet, and provide a much lower one. Through it drill a hole, the same size as that in the centre of the boring collar. The centre of this hole must be in the same line with the centre of the mandrel. The collar, when attached to it, faces the mandrel, and is held firm by a screw. When fixed, the centre hole is opposite the axis of the mandrel; and when the largest hole is used, it clears the top of the headstock to which it is affixed. The end of the work to be bored being placed in the hole which fits it, the tool is held upon the rest against its centre, and the boring is easily and accurately performed.

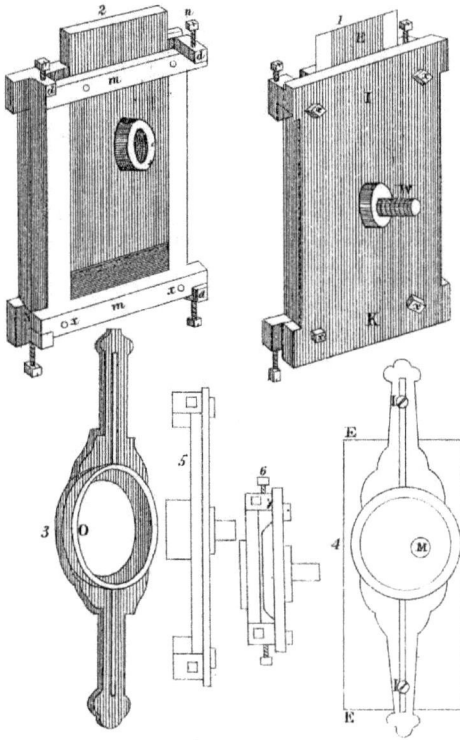

Plate 3.

The collar, D, is perhaps, a simpler apparatus for supporting long pieces of slender wood or ivory, and has this advantage, that the workman can easily make it for himself. It should be constructed of very hard, well-seasoned wood. The foot, E, must be in breadth exactly the size of the aperture in the bed of the lathe, and is kept quite steady by a bit of wood thrust through it underneath the lathe. Into this collar, fit many pieces of wood with different sized holes bored through them, (all exactly in a line with the axis of the mandrel,) so as to admit large or small pieces of work, the sides being grooved, and the supports sawn to correspond: they all slide in with great ease, and are kept quite steady by an iron pin which runs through the top. The fig. G, shows the collar, with one of the supports slipped in, the other sliding upon it; H is the collar, showing the groove, and L L are two supports that fit into it. The *middle* of the apertures of these supports, whether large or small, must always be exactly on a line with the axis of the mandrel, therefore, after they are bored quite true on the lathe, they are sawn exactly across the hole. This is a very useful addition to the lathe, not only as a support when boring holes, but also to be used with the right hand puppet, to give strength to any long slender piece of work, as a screen-handle or a pen-holder; for, being of great length, the stress necessary to the proper management of the tools would be apt to break the wood or ivory, and it is easy to make one of the ornamental mouldings of a proper size, to enable it to run smoothly in one of the supports; if too tight, the wheel will not turn, and if too loose, the work will jerk up and down.

For boring, there are many shaped tools of various sizes; 16, 17, pl. 1, are drawings of the most useful. They have no handles, but at the smaller end a hole is drilled, to admit the point, J, of the puppet, G, pl. 1. Having, with one of the turning tools, made an aperture in the work sufficiently large to allow the boring tool to enter, screw the puppet, G, firmly to the bed of the lathe, then turn the small wheel, K, till the point enters the hole in the tool, which must be steadied by holding it straight and firm with a pair of pincers. Make the wheel turn rather quickly, and with the left hand keep moving the left wheel, K, very gently, so as to force the tool into the wood. After a few turns, stop the lathe, and take out the boring tool, to get rid of the shavings and dust; move the puppet nearer and begin again. A little difficulty will be found in making the tool enter the wood, or *bite*, as it is called, but, by humoring it gently, it will soon take hold; care must be taken to keep it quite straight, and not to go too fast, or it will be liable to break. This method of boring is only used for small hollow tubes, needle-cases, crochet-needles, handles, and small work.

OF WOODS.

There are many beautiful English woods which are excellent for turning; beech is very universally used, and it should be cut into moderate sized pieces and boiled, to render it more durable, and to make it work smoothly.

Elm and chestnut are also much admired; if the latter be dipped in alum water, then brushed over with a hot decoction of logwood, afterwards with one of Brazil wood, it will be made to imitate mahogany. Green wood should never be used, as it is apt to split; it should be kept for at least a year before attempting to turn it. Some persons, if they fear the wood has not been sufficiently seasoned, cut it up and put it in a vessel filled with a ley made of wood ashes. In this it must be boiled for an hour, and allowed to remain in the liquor till quite cold, afterwards it must be dried in the shade.

Old walnut wood is very beautiful; to improve its colour, it may be put in the oven, and when worked, polish it with its own oil, very hot.

Sycamore, when grown in favourable situations, is as white and nearly as hard as holly; the cherry, yew, laburnum, and pear-tree woods, are also very beautiful; but though invaluable for plain turning, they are not hard enough, or of a sufficiently close and fine grain, to admit of ornamenting them in delicate and minute patterns. The milling tools are generally employed for them. Of all English woods, the holly is the whitest, and is rendered still more so by boiling; it is, when very good, used for inlaying, in imitation of ivory. Box is the hardest and toughest of our woods; when cut plank-wise, it is apt to warp, if not well seasoned; but its yellow colour, if highly polished, is much admired, and it will receive the most delicate patterns; it is also used as a substitute for ivory.

The foreign woods are those most prized by the ornamental turner, on account of their hardness, and the beautiful polish which can be given to them. Cocoas, or the wood of the palm, is much used for all kinds of ornamental work. It is of a beautiful brown, streaked with darker veins, and is found in the West Indies.

Ring wood is extremely hard, of a chocolate brown, with black veins; it is a good wood for turning, and comes from Brazil.

Partridge and leopard woods, tulip and snake woods, are also frequently used; the latter is of a very deep red, and very hard.

Calamander wood, a tree growing in the island of Ceylon, is very hard and heavy, and the veins in it most beautifully shaded. The principal colours are a fine chocolate, sometimes deepening almost into black, then gradually shading into a cream colour. It is a very hard wood, and takes a high polish.

African thorn is of a beautiful dark colour, and much prized when it can be obtained good, which is rarely the case.

Ebony, an exceedingly hard, smooth, foreign wood, is much admired by turners. The best is a jet black, free from any veins, and receiving a very high polish. There is some difficulty in keeping woods to prevent them from warping or cracking. The foreign woods, particularly, being usually very dry, often open in fissures while working. The best method of preserving them is to place them in a cool and rather damp place, and to rub a little oil now and then over the outsides, to keep them moist. If the wood is sufficiently large to allow of its being quartered, the danger of its splitting is much less; but the foreign woods are rarely large, as the trees are generally very high, but small in circumference. If foreign woods cannot easily be obtained, box wood and holly may be stained so as greatly to resemble them. The dying woods to be used must be in small chips or raspings. When the wood is ordered to be brushed over several times with the fluid, it should be dried between each time. If the stain is wished to be very deep, the wood should be boiled in the stain.

TO STAIN WOOD RED.

Mix two ounces of Brazil wood, and two of potash, in a quart of water; let them remain in a warm place for some days, stirring them occasionally. With this boiling liquid, brush over the wood till it becomes of the requisite colour; then dissolve two ounces of alum in a quart of water, and, while the wood is wet, brush it over with it. For a pink or a rose red, use double the quantity of potash.

A YELLOW STAIN.

Steep one ounce of turmeric in a pint of spirits of wine; let it stand for several days. Brush the wood over with it. A red yellow is made by adding to the above a little gum tragacanth.

A BLACK STAIN.

Brush the wood with a hot decoction of logwood, then with common ink.

A PURPLE STAIN.

Boil one ounce of logwood and two drachms of Brazil wood in a quart of water, over a moderate fire. When one-half is evaporated, strain it, and brush the wood over with it. When dry, brush it over with a solution composed of a dram of pearl-ash in a pint of water.

A MAHOGANY STAIN.

For a light stain, mix two ounces of madder and one of fustic in a quart of water, and boil them all together; a darker stain is made by using half an ounce of logwood in the place of the madder, and then brushing the wood with a weak solution of potash.

All hard woods are easily polished; first, they are made perfectly smooth and even, with the turning tools, after which rub them with sand paper, then with Dutch rushes, which, to prevent their breaking into small pieces, should be steeped in water. While using these, make the lathe turn quickly round, sometimes one way, sometimes the other, to prevent any unevenness, and keep moving the sand paper, &c. &c., or the edges are apt to cut lines. When this is done, and the work looks smooth, rub over it a drop or two of olive oil, wipe it clean with its own shavings, take it off the lathe, and brush it with a very hard brush, the same as those used for blacking leather.

As it is impossible to turn well unless your tools are in good order, great care must be taken to keep them very sharp. If a bit of the steel splits away, as is frequently the case in turning hard woods, grind it down till it becomes even again, then rub it on the Turkey stone, with a little olive oil, till the edge is so sharp that you cannot see it; for the gouges and hollow tools, thick Turkey stones, rounded at the edges, are sold, which enter into the groove, and the outer edges are rubbed on the flat stone.

As soon as the turner becomes quite master of his tools, he will find ivory much pleasanter to work than wood; it is not so liable to split, it turns smoother, polishes with less trouble, and shows any ornamental work much better than wood. But then it is very expensive, and very difficult to obtain good and white. New ivory may be bleached by exposing it in the sunshine, and wetting it constantly, or it will crack; but till the ivory is cut up, you cannot tell whether it will be good or not. Pieces may be bought cheaper that have a hollow in the middle, they serve very well for pedestals of vases, by screwing another bit into the hole, or for boxes, by gluing in a piece of ebony to fit the hollow. Ivory is polished *before* ornamenting, with putty powder, ground very fine, rubbed on with a piece of linen dipped in water, dry it, and rub very hard with a bit of felt, and the polish will be beautiful. But after it is ornamented, polish only with a brush dipped in water and chalk, or even in plain water.

For dying ivory, it is first necessary to cleanse it from the grease which it always contains, more or less, and which would prevent its receiving the stain or dye. For this purpose, mix half a pound of nitre in an equal quantity of water, tie a string round the pieces of ivory, and dip them in while the liquor is hot, then plunge them into cold water.

TO DIE IVORY RED.

Take half a pound of pieces of scarlet cloth, put them into a clean earthen pot; add one ounce of soft soap, after which pour in three quarts of soft water. Boil all together for half an hour, stirring it frequently, and squeeze the cloth several times, to extract the colour. When this is done, have ready an earthen vessel, put into it as much pulverized alum as will lie upon a sixpence, pour the scarlet liquid over it, and extract all the colour from the cloth by pressing it in a canvas bag. Steep the ivory in this liquor till it becomes of the proper scarlet.

BLACK DYE.

Boil a quarter of a pound of logwood shavings in a quart of water, in an earthen vessel, for half an hour. Steep the ivory in it.

On taking the ivory out of these boiling liquors, immerse it instantly in cold water, to prevent its cracking.

ELLIPTIC TURNING.

This machine is frequently called, by those who do not understand the ellipse, an oval chuck; but it is not oval, for an oval expresses an object that is smaller at one end than at the other. Fig. 1, plate 3, is a front view of the machine. I K is the iron plate to which all the parts (except the ring, hereafter described) are fastened. A screw, similar to that on the nose of the mandrel, is riveted to this plate, fig. W, and upon it whatever you wish to turn is fixed. Fig. 2 exhibits a back view of this machine; at each of the four corners there is a short square pillar, marked D. Within these are two narrow ribs of steel, reaching the whole length of the plate I K. Each of them, being bevelled, forms an angular groove, reaching all its length. By means of these grooves the slider, F, moves up and down.

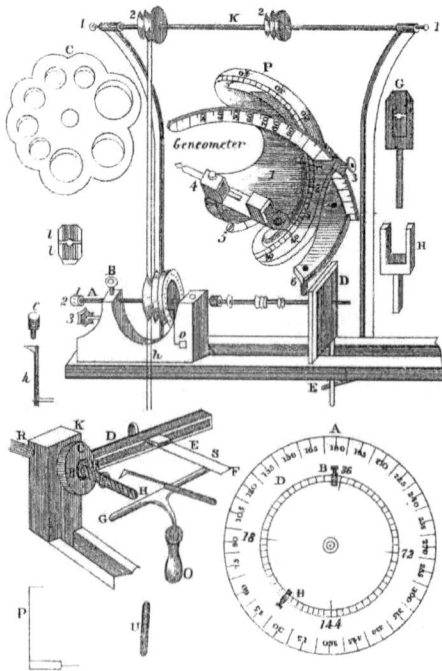

Plate 4.

When the slider is in its place, two pieces of steel, M M, bear upon the side pieces, to which, and to the plate, I K, they are firmly attached by four screws, X X X X. The plate, F, being cast in the same piece as the slide, cannot be thrown out of its place, but moves in a longitudinal direction only, the nut,

L, acting as a stop to prevent its going too far. The space between the end pieces, M M, is just equal to the diameter of the ring, O, Fig. 3, upon the outside of which they revolve when the nut is screwed upon the mandrel. Two arms, R R, are connected with it, and in each there is a groove extending nearly their whole length. This machine is connected with the lathe, and its motion obtained as follows:—E E, fig. 4, represents a headstock, through which two holes are drilled, the centres of which are precisely in a line with the centre of the mandrel, M; the ring is fastened to the headstock by two screws, I I, the shanks of which pass through the grooves and through the holes in the headstock. When the ring is in this position, it will be perceived that it can only move from side to side, and its centre must always be in the same horizontal line with that of the mandrel. Now screw upon the mandrel the nut, L, fig. 2. The plate, I K, if set exactly opposite the mandrel, will revolve in a circle; but if the centre of the ring be the least in the world on one side of the mandrel, it will revolve in an ellipse. When, therefore, the work is fastened to the screw, W, it is quite as easy to turn an ellipse as on the ordinary lathe to turn a cylinder.

Of course, the slider must move with great steadiness and freedom, to effect which, very great accuracy in the workmanship must be observed. Figs. 5 and 6 are two different views of the machine,—5 is the side, 6 the end view.

To turn a hollow sphere, the convex surface is first turned, and perfectly smoothed and finished; it must then be bored with a centrebit, to make an aperture sufficiently large to admit the tool, fig. 7, plate 1, with which the interior must be worked away. As, however, it would require a large aperture to enable you to hollow out the whole sphere, it is preferable to make six openings with the centrebit, each in a line with the centre; they must also be made at equal distances from each other, and every hole must be at right angles with all the rest, except one, which is exactly opposite to it. Place the sphere in a chuck, with the middle of any two of the holes in a line with the axis of the mandrel; turn out as much as you can of the first hole, then bring the other holes forward and do likewise; at last the excavations will be cut through.

To turn the Chinese balls, which are so much admired for their beauty and curious workmanship, we are told to proceed as follows:—As they are composed of spheres one within the other, the holes must be just deep enough to leave the thickness between each little more than the diameter of the smallest sphere. The work must be begun by forming the innermost sphere, and afterwards it must be continued regularly on to the larger, till at last the outer sphere is completed.

An immense, almost an endless, variety of figures may be worked in the lathe by carefully regulating the tools. Beautiful flowers in ivory, equal to the

Chinese carving, are formed by the experienced turner, with small wheels and other instruments: medallions, even, are executed in the lathe, with machinery so constructed that the tool follows on the wood the exact lines of the head that is being copied.

Gold and silver ornaments, such as watches, snuff-boxes, and other trinkets, are worked with what is called a rose engine; plates with patterns indented upon them are fastened upon the mandrel: the screw regulates the tool, which produces an exact counterpart of the pattern.

Many of the copper cylinders used in printing calicoes afford curious specimens of engraving in the lathe. It is impossible to imagine any thing more beautiful than the effect produced, and a whole web of linen is printed by them in a very few minutes. The methods employed in this kind of turning by the various artists who practice it are but little known; and, indeed, to look at the patterns produced, it would be supposed that the graver must have been used to form them. A general idea of the nature of the work is all we are enabled to give. The pattern intended for the cylinder must be cut upon a small steel wheel, which revolves upon an axis. This wheel is then to be held against the copper cylinder, which, quickly revolving in the lathe, carries it round, and receives from it the impression of its pattern. The wheel operates like a punch, but the roughness it makes on the edge of the work is easily polished down.

Having now finished the directions for concentric and elliptic turning, we will, before beginning the ornamental parts of this interesting art, beg to impress upon the minds of our readers that they must never be discouraged by failure in their first attempts, even though they may be subjected to many disappointments; for, as the Bard of Avon expresses it:—

"Oft expectation fails, and most oft there

Where most it promises; and oft it hits

Where hope is coldest, and despair most sits."

ECCENTRIC TURNING.

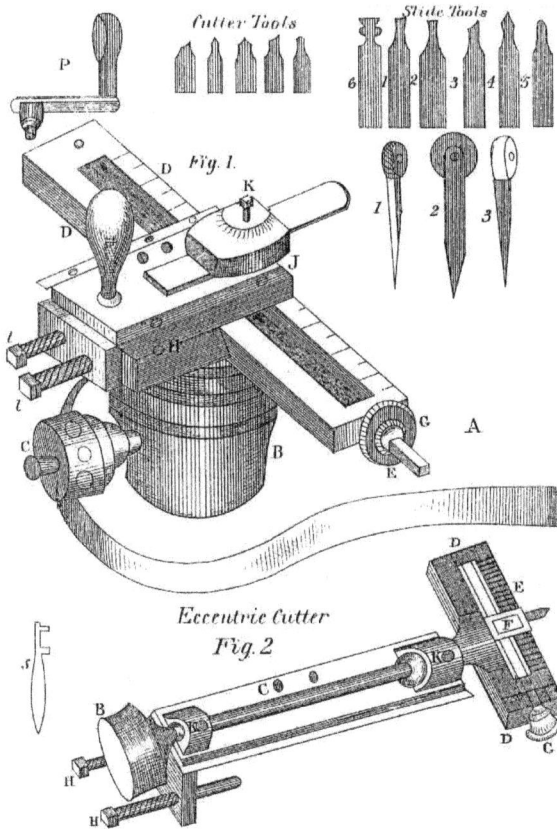

Cutter Tools

Slide Tools

6 1 2 3 4 5

1 2 3

Fig. 1.

Eccentric Cutter
Fig. 2

Plate 5.

This name includes all the various, beautiful, and intricate work for which the powers of a lathe are so justly celebrated, and which, once seen, must be admired by all who love the fine arts and examine the powers of machinery. The eccentric cutter, the drill, the eccentric chuck, and the universal, or vertical cutter, are all indispensable for those who wish to perform ornamental turning, and with them the most delicate and intricate patterns can be worked with a precision and accuracy that are truly wonderful. A sliding, or parallel, rest is absolutely necessary for turning patterns, as upon it, in some degree, depends their exactness. Fig. 1, plate 5, is the drawing of one. The foot, A, screws firmly to the bed of the lathe, like the common rest, and the upper part fits into the socket, B, and, by means of the nut, C, is turned in any direction you wish. D D is a bed of steel, five or six inches long; through the groove down the middle is a screw which passes from one end to the other, and, by turning the squarehead, E, the tool slider, J, is pushed backwards or forwards, while the rest itself remains stationary. The steel bed is graduated. Between the bed and the screw, E, is a small brass wheel, divided into numbers, G; this regulates the position of the tool. The tool slider, J, fits into the bed, H, and slides in and out; J is the part where the tool slips in, and it is kept firm by a screw, K. The handle draws it to and from the work, and the nuts *l l*, regulate the depths of the cuts. The tools belonging to this rest, 1, 2, 3, 4, 5, are of steel, about two inches long, and are ground to different angles: there are various sizes of each. When you intend to ornament a piece of wood or ivory, first turn it quite round, and smooth it with the common tools; you must then make the surface *perfectly flat*, and to accomplish this, adjust the sliding rest at a proper distance, put in one of the round-ended tools, No. 5, let it just *touch* the wood by setting the screws, *l l*; put the lathe in motion, and gently approach the tool to the work; if it cuts too deep, tighten the screws a little more; if too little, push the screw forward. Keep the tool steadily up to the work by pressing on the lever, S, which impels the tool to slide forward, then, with the right hand, turn the head, E, by means of a handle, P, which will thus enable the tool slide to run from one end of the brass bed to the other, at the same time that the lathe is turning rapidly round, and thus the surface of the wood is rendered perfectly smooth, level, and polished. Sometimes, however, it will require to be worked over in this manner three or four times before it becomes quite flat; and frequently, even when it appears level, there will still be a hollow in the middle, which will quite destroy the accuracy of the patterns: to discover whether this is the case or not, take the T square, see plate 1, fig. R; place it with its edge against the wood: if you can see light between it and the work, the surface cannot be even, and requires smoothing over again. The next process is to cut very delicate circles all over the work, at regular distances: this is done to take off the bright look of the wood, that the patterns may appear to greater advantage. Put into the tool slide a double angular tool, No. 4; adjust it so as to make a distinct, but still delicate, cut into the wood. Make one cut in the middle of the work; move the brass wheel, G, one number, or from 1 to 2; make a second cut; move to 3, then to 4, and so on: each circle you will find enlarges gradually, till you arrive at the end of the wood. Pattern 1, plate 6, is a specimen of this work; but remember, it is merely a preparation for other patterns. In doing *side work*, as the side of a box, or a knitting-case, unscrew the screw C, and turn the brass bed of the rest round till in a line with it, then proceed as directed for face work. The wood being ready, we will now go on to give a description of the eccentric cutter, plate 5, fig. 2. Like the tool slider, it fits into the sliding rest, but now it is no longer the work which moves round and the tool stationary, but the wood remains firmly fixed while the tool rapidly revolves and cuts the patterns. For this purpose, it is obviously necessary that the fly wheel should turn the cutter, while the small wheel remains immovable. Several methods are used to perform this: the one given plate 4, fig. K, is the easiest. The frame here represented should be of iron, firmly screwed to the bench of the lathe, and of sufficient height to be about a foot above the head of the workman. In front is a spindle, which works in two nuts, No. 1 1, exactly in a line with the mandrel. Two wooden wheels, No. 2 2, are fastened to this spindle; the one on the left hand remains stationary over the fly wheel of the lathe, by which it is turned, the other slips backwards and forwards, according to the work it is required to do. Take off the usual catgut from the fly-wheel, and pass a long one over it, and over the small wheel on the over-head frame, No. 2. When the cutter, 2, plate 5, is fixed in the slide rest, draw the other small wheel on the spindle, No. 2, forward, till just above the rest, then pass a catgut over it, and round the small brass wheel, B, of the cutter, and the whole will turn together. The brass wheel of the lathe must then be fixed at one particular number. It is usually divided into three hundred and sixty divisions, each marked by a small hole in the brass wheel, as in fig. A, plate 4, and it is by properly dividing the numbers on this plate that the accuracy of the patterns depends; to keep the wheel steady, a small steel key, h, plate 4, is slipped into the brass knob, O, plate 4, and the other end, being pointed, enters into one of the small holes, say the one marked 360; the work is then immovable until you remove the key into another hole. The cutter itself is of brass, with a spindle, C, which works in two brass collars. At one end is the wheel, B, by which it is turned, at the other a steel frame, T D, which is marked on the upper edge in small lines, E, to regulate the quantity of eccentricity. The steel tool-box, F, holds the tool, which is kept firm by a small screw underneath. By means of a screw through the frame, D D, similar to that of the sliding rest, the tool is pushed backwards and forwards, and cuts a large or a small circle. G is the nut that moves it, and it also is divided into numbers. This cutter, for many patterns, is quite as useful as the eccentric chuck, but, in conjunction with the former, is invaluable, and the patterns performed by them may be multiplied according to the taste and genius of the turner. The two screws, H H, fix the depth of the cut; the wheel of the sliding rest determines the necessary distance, that of the cutter the eccentricity, while the brass wheel keeps the pattern accurate. The tools, Nos. 1, 2, 3, 4, 5, plate 5, are of various forms and sizes: care must be taken not to break them, and the cutter must be constantly well oiled with olive oil; the holes, K K, are made to receive it. In using this tool, make the wheel of the lathe go very quick, but approach the tool very gently and slowly to the work. The better to do this, the lever, S, is used; it enters into one of the holes in the side of the cutter, and into a similar one in the sliding rest, so that by a slight pressure the tool is impelled gently forward. There are generally about two dozen tools shaped like those in the plate, but of various sizes.

To imitate the second pattern, plate 6, which, it will be perceived, is a number of circles, slide the cutter towards the edge of the work by turning the screw of the slide rest, then, to make the outer circle of the border, turn the small nut, G, of the cutter six times round; put a double angular tool into the tool box, screw it firmly, adjust your cutter by the screws, H H, to the proper distance, and stop your wheel with the steel key at 360. Cut one circle, alter your wheel thirty numbers, cut another circle, then thirty numbers more, till the twelve circles are all cut. This done, alter the wheel fifteen numbers, which will make the circle cut half through two of the former ones, then move 30, as before, till the twelve are done. For the small inner circles turn the nut, G, of the cutter two turns back, to reduce the size, and move 30 as before. The pattern in the middle is still circles, though differently arranged. Draw back the cutter till a little past the middle of the work, then unscrew the nut, G, fifteen turns, cut one circle, move the brass wheel 15, cut another; move 15, cut another; then move 45, cut another; move 15, then 15, and so on till all are finished.

Third pattern.—Set the cutter at five turns, and move five numbers on the brass wheel for every circle. For the middle pattern, turn the nut of the cutter eight times, move ninety numbers on the wheel, till four circles are cut; then move 45, cut another; then 90 again. To make the small circles, turn the nut, G, backward two turns, and move 90 as before, for the four first circles; then move 45; then 90 to the end.

Plate 6.

Fourth.—The border of this pattern is the same as the one in No. 3, only you place a piece of wood about a quarter of an inch in depth *across* the bed of the lathe, letting it pass half under the slide rest; screw the rest down firmly, and proceed as above. By thus raising the rest in a slanting position, half the circle only is cut, which has a very pleasing effect. For the centre pattern, set the cutter at a small circle, cut one in the middle, having the wheel fixed at 360. Then move the brass wheel of the slide rest, plate 5, three numbers for each circle; these done, return to the middle circle and continue the same to the other end. Then move the brass wheel of the lathe to 45, and repeat as above, three numbers for each circle, till the next row is finished; then again move forty-five numbers on the large wheel; and so on to the end.

Fifth pattern.—Set the cutter at a large circle, the brass wheel at 360. Move 5 for four circles, then move 15, then 4 four times more, then 15 again, and so on to the end. For the middle, set the cutter to a small circle, and move the brass wheel of the lathe 90 four times.

Sixth pattern.—Set the cutter to the largest circle it will make, begin in the middle of the work, set the wheel at 360, and move ten every time.

To assist the beginner, a drawing is given of the wheel, supposing the largest number to be 360, A, plate 4. By this the numbers are all arranged; a table of the divisions is also given; for on the proper and accurate calculation of the numbers depends the exactness of the patterns. The plate on the wheel divided at 144 is the plate for marking half-circles and arcs, as in plate 12, fig. 3, which shall be described hereafter. These patterns are very beautiful, and, from the arcs being gradually reduced in size, they have a curious and elegant appearance, but they cannot be worked *without* an eccentric chuck.

THE DRILL.

The drill is a most useful auxiliary to the eccentric cutter, not only for drilling holes, which it does with great nicety and speed, but also for making mouldings and patterns of various kinds. Unlike the cutter, which moves either in a large or small circle, the drill can only work upon its own centre, and therefore the size of the pattern depends upon the tool placed in it, its position being regulated by the screw in the sliding rest, into which it slips like the cutter. It is turned by a rope exactly the same as in the directions already given for the eccentric cutter, and greatly resembles it in shape, except at the end, A, plate 7, which is made just to receive the tool, and a small screw keeps it firm. Suppose you wish to ornament anything—say, the pen-holder, plate 9—with concave mouldings, as in A. Having set the rest, by the aid of the T square, exactly in the same slanting direction, put a round-ended tool, No. 3, into the drill, set it to the proper distance, fix the brass wheel at 360, cut a round hole; move the wheel to 72, cut another, and so on till five are drilled; then, without altering the rest, put in a smaller round-ended drill, hold it by means of the lever well up to the work, make the large wheel go very quick, and slowly turn the screw of the slide rest, so as to impel the drill which ever way you wish. Do not cut too deep at first, or you will break the tool; if a great depth is required, go over in the same line three or four times. This done, count 72, as before, and proceed the same; the pattern will have the appearance of the pen-holder in the drawing. The end, B, after being turned to the shape, is ornamented in small holes, that resemble a honeycomb, in the same manner, with a round-ended drill; counting so as to make them fit nicely between each other. To do this, you must be able to subdivide your first numbers. Thus, suppose you drill a round hole at 1, 40, 80, and so on till you come to 360, these numbers can be divided by beginning the next row (which must be begun the breadth of the tool from the former one, by turning the slide-rest screw half, or a whole turn, according to the breadth of the tool) at 20, then count forty numbers as before; but if you had taken forty-five instead of forty for your number, you could not have divided it evenly in the second row, so as to make the holes intersect each other. Be very careful to remember how many times you turn the slide-rest wheel, that each cut may be of equal length. There are generally about four dozen tools belonging to the drill, of various sizes, but of the shapes given in plate 7. Nos. 1 and 2 drill large or small holes, as for instance, round the sides of a turned pincushion, or needle-book, for the stitches to go through; 3 is used to make concave mouldings, or to cut quite through the work in straight lines, as in the lighter case, E, plate 9. This pattern looks very pretty, and is quite easy, if your work is turned sufficiently thin; it should be lined with coloured velvet. No. 4 makes concave mouldings flat at the bottom; these tools are also used to cut round dots: 5, cuts small or large beadings, which give great lightness and finish to the work. Having made a moulding with the hollow tool No. 2 of the slide rest, choose a beading tool that just fits the moulding, put it in your drill, set it to cut sufficiently deep to be quite round at the top; having cut one, count by the brass wheel the proper distance, to make them fit close, but without one spoiling the shape of the other; the round dots on the bottom of the lighter case are intended to represent these beads. Tool No. 6 cuts mouldings of the same shape as the drawing. In using this and the tools Nos. 7 and 8, be careful, after the first cut, where you place them for the second, to make them fit; and in using all the drill tools, make the lathe go as quick as you can, but move your *tool* very *slowly*, and keep the drill slide well oiled.

LIST OF NUMBERS ON THE SMALL WHEEL AND ECCENTRIC CHUCK.

	LATHE WHEEL.					CHUCK WHEEL.		
No. of divisions for one cut.	No. of cuts to complete the circle.	Odd numbers.	No. of divisions for one cut.	No. of cuts to complete the circle.	Odd numbers.	No. of divisions for one cut.	No. of cuts to complete the circle.	Odd numbers.
1	360	—	1	96	—	1	120	—
2	180	—	2	48	—	2	60	—
3	120	—	3	32	—	3	40	—
4	90	—	4	24	—	4	30	—
5	72	—	5	19	1	5	24	—
6	60	—	6	16	—	6	20	—
7	51	3	7	13	5	7	17	1
8	45	—	8	12	—	8	15	—
9	40	—	9	10	6	9	13	3
10	36	—	10	9	6	10	12	—
11	32	8	11	8	8	11	10	10
12	30	—	12	8	—	12	10	—
13	27	9	13	7	5	13	9	3
14	25	10	14	6	12	14	8	8
15	24	—	15	6	6	15	8	—
16	22	8	16	6	—	16	7	8
17	21	3	17	5	11	17	7	1
18	20	—	18	5	6	18	6	12
19	18	18	19	5	1	19	6	6
20	18	—	20	4	16	20	6	—
21	17	3	21	4	12	21	5	15
22	16	8	22	4	8	22	5	10
23	15	15	23	4	4	23	5	5
24	15	—	24	4	—	24	5	—
25	14	10	25	3	21	25	4	20
26	13	22	26	3	18	26	4	16
27	13	9	27	3	15	27	4	12
28	12	24	28	3	12	28	4	8
29	12	12	29	3	9	29	4	4
30	12	—	30	3	6	30	4	—

31	11	19	31	3	3	31	3	27
32	11	8	32	3	—	32	3	24
33	10	30	33	2	30	33	3	21
34	10	20	34	2	28	34	3	18
35	10	10	35	2	26	35	3	15
36	10	—	36	2	24	36	3	12

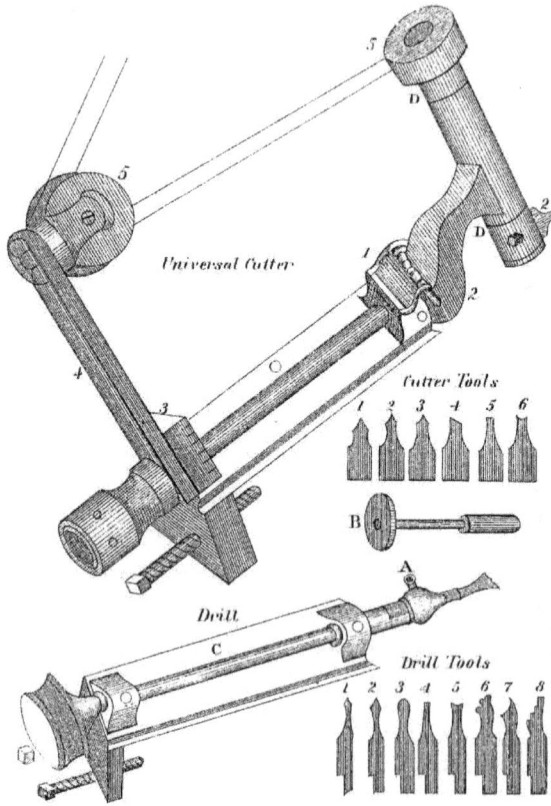

Plate 7.

These tools are also useful for cutting out the edges of work in vandykes, as in the top of the lighter case, E, for which pattern the tool No. 8 was used, the gimped edge being cut with No. 2; the end of the needle-case, D, in plate 9, is done with a round-ended drill, No. 3, in the same manner as directed for the end of the pen-holder. In the patterns Nos. 1 and 6 of plate 11, the straight lines near the centre are cut with the round-ended drill.

GONEOMETER.

All the eccentric tools require the greatest care in sharpening, and the above ingenious machine has been invented for this purpose; it is represented at P, in plate 4. The upper part, 1, is a plate of brass, the outer edge, 2, is graduated as high as 50 each way, beginning at the tongue, 3. Beyond the numbers is a groove, in which one end of the tool slide, 4, slips, and is firmly fixed (so as to point to any of the numbers that suit the angle of the tool) by a nut underneath. The whole plate, 1, is raised and lowered at pleasure by a small hinge at 5, and the requisite height is fixed and settled by counting the numbers on the steel tongue, 3. Underneath is another brass bed with two feet, upon which the machine rests, while the front part leans forwards and rests upon the end, 5. The box that contains the goneometer has three drawers in it; the first is lined with brass, except for about three inches in width, which space is covered with fine Turkey stone. To use this machine, take out this drawer, lay the feet of the goneometer on the brass, put the slide-rest tool, No. 4, into the tool box, screw it firm, then slide the tool box along the groove till it arrives at the proper angle, say 45; fix it with the nut underneath, then raise the whole brass plate, 1, sliding it along the tongue, 3, till at the exact height necessary for the tool to touch the Turkey stone. Rub it backwards and forwards upon it with oil. When one side is sharp, move the tool slide to the opposite angle, to sharpen the other. The small eccentric tools are placed in the steel case, U, and the case fits into the tool slide.

CIRCULAR REST.

The rest we have already described, called the Parallel Rest, works, as its name sufficiently expresses, in a straight line; it is therefore useless for ornamenting spherical objects, and the circular rest has lately been invented to supply this deficiency. It is a most ingenious contrivance, and perfectly fulfils the purpose for which it was invented.

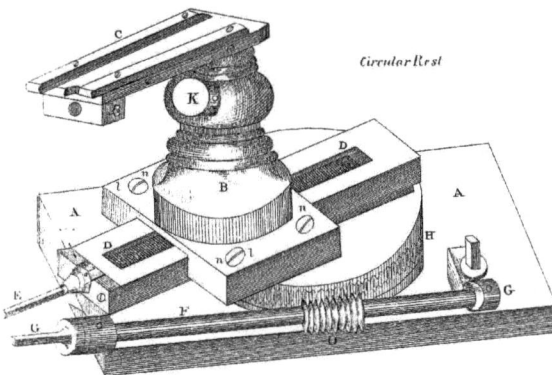

Circular Rest

Plate 8.

The bed, A A, is the same as in the parallel rest, and screws in the same way to the bed of the lathe. B is a brass socket and pillar, which support the bed C, in which the tool box slides; they are formed in the same piece with the lower bed, L L, and firmly fixed into the lower part, which is grooved, by four brass-headed screws, N N. The grooves enable the whole socket and tool box to slide backwards and forwards on the bed of steel, D D, and by means of the screw which passes through it the workman regulates the advance or retreat of the rest to or from the work in a *straight line*, while the lower part remains stationary. The small wheel, E, is graduated, and turned by a key, to enable the turner to count the distance. F is a steel spindle, which works in two brass collars, G G; about half-way down. The spindle is formed into a screw, O, which turns upon the brass wheel, H, and by moving the nut, J, moves the whole rest in a circular direction, in the same way that the screw in the steel bed, D D, impels it in a straight line: by this ingenious contrivance all objects that are round or spherical can be ornamented: such as balls, the globular sides of vases, or small baskets, in every variety of pattern. The learner will easily discover the proper method of using this rest, which is very simple in its mechanism and use. The nut, K, is for setting the tool slider at different angles, the same as in the parallel or sliding rest; the tools to be employed are those of the above-mentioned rest.

THE VERTICAL, OR UNIVERSAL CUTTER.

PLATE 7.

This cutter, which also fits into the slide rest, is different from the others; and, as its name denominates, the patterns it can cut are almost endless, for it may be turned in any direction. The bed of the slide is the same as with the drill; at the end, No. 1, is a screw, which, when turned by the key, B, inclines the tool-holder, 2, to any angle that may be required, and it is regulated by the lines on the brass plate, 3, which are marked by a small steel point; so that if you wish to cut out a pattern slanting to the right, and another to correspond slanting to the left, you have only to mark the number on the plate, where you cut the first, and then with the key move the cutter to the same position on the opposite side. When the tool-holder, 2, stands straight, as in the plate, the tool cuts horizontally: when it is screwed down to the last line on the plate, it cuts perpendicularly, but the cut always scoops out; and by putting the tool as far out of the holder as you can, the cut will be larger, and the scoop deeper. The back support of the pulley, 4, moves with the tool-holder; and the pulleys, 5, correspond with each; the back one is turned in the same direction as the front one, by unloosing the screw, 6. The gut, after passing over the pulley on the over-head frame, comes through the two back pulleys of the cutter and round the front one, as in the plate; but when the cutter is screwed flat, a short cord, the same as that used for the drill, is sufficient; and the back pulleys are then not necessary. This tool requires constant oiling at D D, for the friction is very great; it also demands much care and delicacy in using it, as it must not be incautiously thrust against the wood, or it will stick fast, and not turn round. There are about four dozen tools, all of the shapes given in the plate, but of various sizes. This cutter slides into the slide rest, and the depth of the cut is, as usual, regulated by the screws at the end, and the cutter is brought to the work by the aid of the same lever as is used for the drill and eccentric cutter. The patterns cut by this tool are counted and regulated by the numbers on the brass wheel, as with the others. When the turner has once tried it, he will readily discover its advantages for all kinds of ornamental work, particularly for the sides of boxes, needle-cases, and many other articles; but it is needless to give many drawings of the patterns, as they much depend upon the taste of the turner.

One beautiful design for a lighter case, or small basket, is worked with this cutter with a flat-ended tool. Turn the work very thin, chuck it firmly, make one cut deep enough to allow the tool *just* to cut through, and no more. Count thirty on the brass wheel of the lathe every time; the next row make the cut deeper, so that the opening will be larger, then move 60. By this means a piece of ivory will be left standing out, with openings cut between, which, when lined with coloured velvet, looks light and elegant. This row count sixty every time; the next, thirty, as at the beginning.

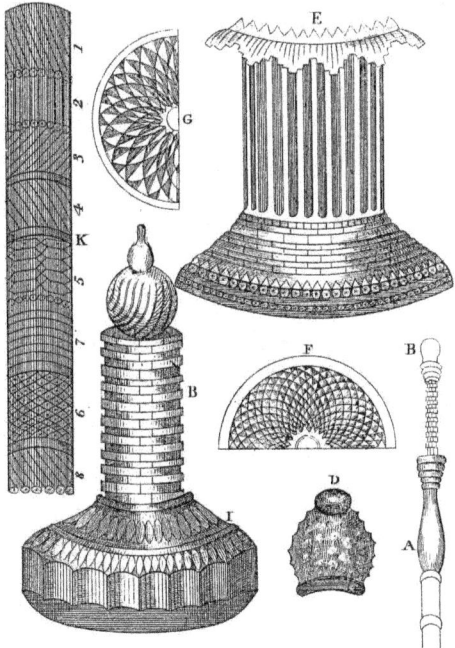

Plate 9.

The needle-case in plate 9, fig. K, is entirely ornamented with this cutter in a variety of patterns. Figs. 1, 3, 4, and 8, are worked with the flat and round-ended tools; 2, with the tool No. 6; 5 with the tool 3; 7, with 4; and 6 with No. 6.

The bottle, B, is also ornamented with the same tools. It is hollow down the neck, and is intended to contain a tin or glass of water; in the stopper is glued a small camel's-hair brush, which rests in the water, and is meant to be used for wetting postage-stamps, and fastening them upon letters. The bottom of the neck unscrews at fig. 1, and the part below it is hollowed out like a box, to hold the stamps. The neck is ornamented in steps. Set the cutter quite flat, facing the side of the work; use the tool No. 5; cut one line, move the brass wheel of the lathe just far enough to make the second cut join the first, and the same all round. In the next row make the first cut half-way between the others; thus, if you have begun at the numbers, 1, 6, 12, in the second row begin at 3, then go to 9, then to 15, and so on. Every row of steps begins half-way through the former ones, and for each row move the cutter on the slide rest the breadth of the tool. The convex moulding at the bottom of the neck is done by putting one of the hollow tools of the sliding rest into a handle, and rounding the ornament with it. The other patterns are all done with the same cutter, by placing the tools at different angles. In the same plate, 9, the two patterns F and G are very beautiful, they are worked with the eccentric cutter. F resembles the scales of fish lying one over another: put the most angular tool, No. 5, (of the cutter tools), into the cutter; describe a circle from the edge of the middle line to the outer one; cut one circle very *deep*, move ten numbers on the brass wheel of the lathe, and so on to the end.

The pattern, G, is worked exactly the same, only a less angular tool, No. 4, is used, and fewer numbers are counted: it resembles leaves, one lying over the other.

The stopper of the bottle, B, is also ornamented with this cutter, and with an angular tool. First turn the stopper quite circular, (the knob at the top must be glued in afterwards). Set the sliding rest at a convenient distance, and place under it one of the slide-rest tools, so as to raise it in an uneven manner; screw the rest firmly, and set the circle just large enough to encircle half the stopper. By this means the tool cuts the UNDER part of the circle, and passes over the other half: this pattern must be cut very deep, and about five numbers be counted between each cut on the brass wheel of the lathe.

D is another pattern for a stopper; it is worked with a round-ended drill. Cut the holes rather deep, and as near as you can to one another, only leaving a very little thin shell of ivory between them, and the pattern will resemble a honeycomb. The lighter case, E, is ornamented with the drill and vertical cutter. Turn the upper part very thin, then drill long lines quite through the ivory, leaving a space between each; this looks light and elegant, if lined with coloured paper. The base is cut in steps with the vertical cutter; count twenty for each cut; the second row, make each cut between the former ones, moving for each row the breadth of the tool. The top is cut out in leaves with the drill tool No. 8.

To sharpen all these tools, use the goneometer, taking care that you fix it exactly at the proper angle, by counting the position by the numbers. If you do not place the tool just at the right angle, it will cut the patterns quite crooked.

ECCENTRIC CHUCK.

"The centre mov'd—a circle straight succeeds;

Another still, and still another spreads."

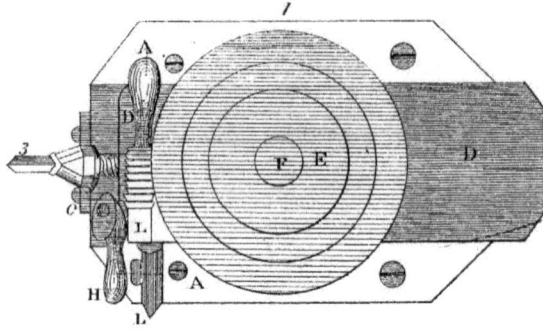

Eccentric Chuck.

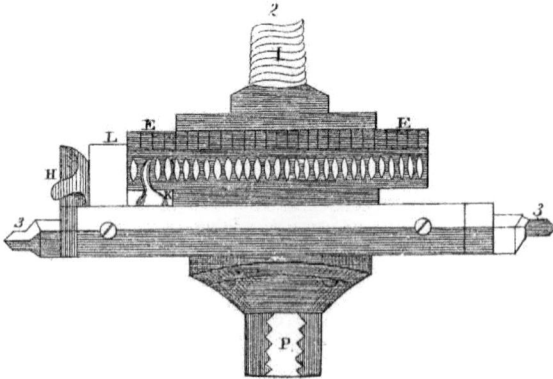

Plate 10.

Perhaps the most useful of all the additions to the lathe is this chuck, for by its aid the turner can alter the centre of his work as he pleases, producing a great variety of circular lines of different sizes, and other ornaments, which cannot be attempted without it. The tools are those already described belonging to the slide rest, which latter, when used with this chuck, assumes quite a different character than when employed with the eccentric and universal cutters and the drill. With these three, it merely serves to mark, by aid of its screw, the distance they require to be from the work, and the position in which they are to be placed, but does not assist in the size or form of the patterns to be cut; with the eccentric chuck, however, it is quite different. The rest marks the *size* of the circles, while the chuck fixes their *position.* Thus: suppose you wish to cut a circle, or series of circles, not as in the pattern, No. 1, plate 6, which begin from the centre and gradually enlarge, but like those in No. 3; the slide rest must be set to the proper position, to enable the tool to cut the requisite circle, and then the chuck must be screwed down till the tool arrives opposite the exact place where the pattern is to be worked. With this chuck the over-head frame is no longer used; the brass and fly-wheels being employed together, as in common turning, the chuck having a wheel of its own by which to regulate the patterns. In the annexed plate, fig. 1, is a front view of the eccentric chuck; fig. 2 is the same seen sideways.

A A are two brass plates, with a screw at the back of them, as shown at P, fig. 2; by this screw it fastens on to the mandrel of the lathe like a common chuck. A A are so shaped as to admit of D sliding up and down them, and the four screws enable the slide to be tightened or loosened. Down the middle of the chuck is a screw with a very fine thread; it is turned at either end by the square head, 3, and thus D slides up and down at pleasure, but cannot in any way get out of its proper position, and, above all, can never shake in its bed. E is a brass wheel divided into 120 teeth, upon which the distances are calculated, the same as on the other brass wheel; in the middle of it, F, is a screw, the same in size as that on the mandrel, upon which the chuck that holds the work must be screwed. The wheel, E, has a spring screw, which, by turning back the spring, H, enables you to move the wheel round as many numbers as you wish, and the steel point, K, marks the one you wish to set it at. L is a small wheel with four numbers marked upon it; by setting it at O, the steel screw passes from line to line of the wheel, E, fig. 2; by turning it one line further, it stops at $\frac{1}{4}$; another line, it stops at $\frac{1}{2}$; and a third, at $\frac{1}{3}$; thus enabling you to cut your patterns as fine and delicate as you like.

The square heads, 3 3, are placed at each end of the chuck, so that the plate D can traverse the whole length, and thus enable you to ornament a square or an oblong piece of work; both heads have four numbers marked upon them, by which you regulate the motions of the chuck. For instance, when D D is screwed up level with the black line O, the screw F will be exactly in the centre of the lathe. Now, suppose the tool in the slide rest is set to cut a very small circle, and you wish, having done that one, to cut several others

of the same size round it, thus, you must turn the chuck down one turn by counting the numbers on the square head, No. 3; and having cut one circle in the proper place, count as many numbers on the brass plate, E, as are necessary to make them fit properly. Should you wish to make a second row of circles larger than the first, besides turning down the chuck, you must, to enlarge them, turn the screw of the slide-rest one turn forwards. All the patterns already given for the eccentric cutter can be done equally well with the eccentric chuck, except the border pattern in No. 4, which can only be worked by the cutter. Patience, calculation, and attention will enable you to perform the most beautiful, minute, and intricate patterns with the eccentric chuck; but though I have endeavoured to give many specimens, when printed they only in a slight degree convey an idea of the real beauty of this kind of turning; the depth of the cut will wholly alter the appearance of a pattern, and the same circles cut with another shaped tool present quite a different appearance; in all, however, there is one rule, which must never be forgotten—viz., be very careful to make your wood or ivory *perfectly* smooth, flat, and even, before attempting to ornament it, or you will be disappointed by finding that your circles are cut deep in one part and are scarcely visible in another; also remember to *line* it, as directed when treating of the slide-rest, for this gives much effect to the work.

Plate 11.

Pattern 1, plate 11. Begin by the shell. Having smoothed and lined your wood, put a double angular tool (No. 4 of the slide-rest tools) into your rest; turn the slide-screw till the tool is exactly in the centre of your work, when, if you move the wheel round, you will find the tool will only cut a dot. The outer circle is the one with which to commence; turn the slide-rest screw forwards ten whole turns, and you will find you have the circle of the proper size. Now approach the tool to the wood, cut the circle carefully at first, till you decide on the proper depth for the cuts; then set your screw guides, and proceed as follows. For the other ten circles, move the eccentric chuck downwards half a turn, and diminish their size by moving the slide-rest screw backwards half a turn, so as to keep the lower part of each circle in the same place. To have a good effect, shells should be *well cut up*—that is, each circle should be of a sufficient depth for the edge of the cut to meet the former one, and thus efface the lines made in preparing the wood. The rays from the centre are worked with the drill; put a fine round-ended tool, No. 3 of the drill tools, into the drill; place it in the slide rest, stop the fly-wheel, arrange the cords, and set the tool with the screw guides to cut very little at first. By means of the slide-rest screw, push the drill forwards till the tool touches the outer edge of the shell; hold it well up to the work with the lever; make the lathe go very quick, and move the drill on the slide forwards very slowly, by turning the slide-rest wheel round with the key 12 turns FORWARDS, then the same backwards. Move 12 numbers on the eccentric chuck wheel for each of the other middle lines, and cut them as above; then move the slide-rest half a turn forwards to shorten the line, and move 1 number on the eccentric chuck wheel, and cut a line moving the slide-rest 11 turns forwards, by which it will be equally shortened at each end. Move to the other side of the long lines and do the same. The whole pattern is done in this manner, counting one for every line on the eccentric chuck wheel, and gradually shortening them by reducing the turns of the slide rest. For the third pattern, which would look better done in dots than in circles, take away the drill, put back the slide rest tool-box, and take a flat-ended tool; set it to cut a dot about the same size as the circles, just deep enough to efface the under lines, so that each dot will look bright and shining. Having cut one, count 2 on the eccentric chuck wheel for each of the other 60; then screw the chuck downwards 2 turns, so that the next row may be just above those already done; cut five dots, counting 2 numbers for each, then pass over four and cut five more; and so on to the end. The upper circle is worked precisely the same as the lower one. The outer pattern is done like the arc patterns in plate 12, counting 5 for each arc on the eccentric chuck.

PATTERN 2.

The middle pattern is formed of eight shells, with only four lines in each; but these parts of shells begin at the centre point, not round it, as in the former pattern. Use a double angular tool, screw the tool slide out 10 turns, and the eccentric chuck *downwards*, till the outer or largest circle just touches the centre; cut one circle, move 15 numbers on the eccentric chuck wheel; cut another, and continue the same till the eight are done; then diminish each as directed for the shell, counting 15 for every circle. The second pattern, of rings one within another, is worked with the same tool; set it to the centre point of the wood, screw it forwards four turns, then screw the chuck downwards until you can cut a circle just above the shells; count 6 numbers on the eccentric chuck wheel for each of the 20: then diminish your circles by turning the slide-rest screw *backwards* half a turn; do the second row, counting 6 as before, and the same for the two inner ones. A still prettier way of doing this pattern is to cut steps instead of circles. Put a fine flat-ended tool into the slide, set it to cut a circle the same circumference as the largest in the pattern; then reduce its size as before, and cut the next deeper, by screwing out the screw guides a very little; the third is still smaller and deeper. A little dot should remain in the middle, standing up as high as the level of the wood; to do this pattern the wood must be thick, as the steps require a certain depth. The outer pattern is done in the same manner as that in pattern 1, only the numbers are counted differently.

PATTERN 3.

Count 40 on the eccentric chuck wheel for each of the three large circles, diminish one turn of the eccentric chuck. Then diminish the circle two turns of the rest screw, and lower the eccentric chuck two turns, and cut a circle every 20 numbers; the others are done the same. For the second pattern of circles, cut one for every number on the eccentric chuck wheel; then move one number on the eccentric chuck wheel, and move the eccentric chuck downwards one turn, and cut a circle every 15; by moving the chuck down, and advancing one number every time, the circles are made to incline sideways. The outer pattern is much the same as in the two former ones.

PATTERN 4.

This is a double shell in the centre. Begin with the large circle, having fixed the eccentric chuck wheel at No. 120. Having cut one whole shell, move the wheel to No. 60, and work the other in the same way. For the circles, keep the tool in the same place on the slide rest, and screw the eccentric chuck down two turns to clear the edge of the shell; cut one circle every 12 numbers on the eccentric chuck wheel, enlarge the circle by moving the slide rest one turn forwards; cut the second row, and so on till the fourth or largest one is done, then diminish the size for the other two WITHOUT altering the chuck, but counting the same. The white dots are worked out with a flat-ended tool, cutting one every six numbers on the eccentric wheel; then move your chuck downwards three turns, and cut one every 12; this done, move back to the former row to the dot that comes between the circles, replace the double angular tool, enlarge your circle, and cut one every 12; screw the chuck downwards three turns, and proceed the same; then screw the chuck upwards one turn and a half, which will bring your tool to a level with the middle of the two former circles; move the eccentric chuck wheel one number on each side of the former ones, and the rosette will be finished. The border is formed of plain circles, omitting 3 numbers every ninth cut.

PATTERN 5.

The middle and second row of circles are too simple to require explanation. For the clusters of rings, begin with the middle row; cut two a little apart from each other, then one through the middle of them, and so on to the end; set the eccentric chuck wheel to the *middle* circle of one of the clusters, and move the eccentric chuck downwards, till the tool can cut through the middle of the others, then upwards the same number of turns. The extra half circle is made by moving the wheel of the lathe with the hand, as for the arc patterns, and counting the numbers on the eccentric chuck wheel, *between* those already formed.

PATTERN 6.

This is merely a circle of shells, worked the same as in pattern 1; only with this difference, that each circle is done to correspond in each shell, instead of working every one separately, which would be more tedious; 15 is counted on the eccentric chuck wheel for each row of circles. The rays in the middle are done with the drill; these can also be done without an eccentric chuck; the numbers must then be counted on the fly-wheel, which, for this kind of line work, is kept steady with the stop.

PATTERN 1, PLATE 12.

Cut a small circle, then one every 12 numbers on the eccentric chuck wheel. Move the chuck downwards two turns, and the eccentric chuck wheel 1 number, either backwards or forwards; if the former, the rows of circles will incline to the right; if the latter, to the left. Proceed the same for every row of circles, always counting twelve between each, and advancing the chuck 1 number at the beginning of every new row.

PATTERN 2.

Begin with a small circle in the middle, then turn the eccentric chuck down two turns, and cut one every 15. For the other two rows proceed the same, lowering the chuck two turns, and counting 15 for each on the eccentric chuck wheel. For the chain work, cut a circle every 15 near the edge; then screw the eccentric chuck one turn upwards, move 1 on the eccentric chuck wheel on each side of the others; thus, if the upper row of circles begin at 120, the second row will begin at Nos. 1 and 119, then at Nos. 14 and 16, and so on: the third and fourth rows are the same as the two first; and the last is cut between the others.

Plate 12.

Patterns 3 and 4 are called arc patterns, and are worked in a different manner. In plate 4, fig. A is a drawing of the brass fly-wheel of the lathe, with the numbers marked upon it. D is a brass circle half an inch wide, which is attached to it: and upon which are marked 144 lines, every 18 of which is described by a number. B B are pieces of brass which, by means of a groove in the edge of the circle, fit upon two steel slides that fit into the groove, and are kept firm by two nuts, which fix them opposite any number you wish. P is a long thin piece of steel, which enters by one end the bed of the lathe, so as always to stand quite upright, by which means, when the brass fly-wheel is gently turned with the hand, the flat end of the stop, P, catches upon the lower piece of brass, B, and if you reverse the motion of the fly-wheel, the other piece, B, will rest upon the stop, so that you will find the wheel can only turn half or three parts round, just as you choose to set the stops. These patterns, it will immediately be seen, are worked with the hand, which must guide the wheel slowly up and down; therefore slip off the cord, and having arranged a double angular tool in the slide rest, hold it up to the centre of the work, put the stop, P, into the hole, *n*, of the bed of the lathe, and gently move the wheel till you find the extent of circle you wish to cut; fasten on one slide firmly, then fix the other end of the arc, and fasten the other slide.

This done, approach the tool carefully to the wood, and cut the first long arc by moving the fly-wheel with your left hand as far up and down as the stop will allow; then move 20 numbers on the eccentric chuck wheel for every one of the other five arcs, and cut them in the same way. After these are completed, move one number on the eccentric chuck wheel and unscrew the UPPER slide, and move it down as many lines as you think fit, in order to shorten the next arc; count 20, as before, for every other of the 5. Each row is worked the same, shortening the arcs and advancing one number on the eccentric chuck wheel for each row. Pattern 4, and the border in pattern 1, plate 11., all are done in the same manner, only counting differently. A great variety of arcs can be worked, and have a very curious effect, but you must be careful in cutting them not to let the tool go too deeply into the wood or ivory, or it will either stick fast, or else cut jagged, uneven arcs; rather prefer to spend more time over the work, by working the arcs over and over again till the proper depth is obtained.

These patterns will enable the learner easily to discover and invent many others, all beautiful when properly and neatly worked.

PATTERN 5.

Cut a circle every 24 numbers of the eccentric chuck wheel. Lower the chuck one turn, and cut a circle on each side of the former ones, so as to form the five knots of threes. The edge is worked like the arc patterns, making each half circle meet and join with a dot, cut with a small flat-ended tool.

PATTERN 6.

For the border cut three circles, one for each number on the eccentric chuck wheel, then cut another at 6, 7, and 8, and so on till all are done. The pattern in the middle needs no explanation.

Plate 13.

Plate 13 contains several specimens of eccentric turning. The temple may be made either of wood or ivory, and be left open, as in the plate, or have a coloured card-board case made to fit inside it. The upper and lower parts of the pillars, six in number, are fluted with a round-ended drill, but the middle parts are worked with the eccentric chuck, and are curious specimens of its powers. They much resemble a spiral staircase, and though at first sight they may appear difficult, yet they are not so when properly explained. Begin by turning the wood or ivory perfectly cylindrical with the slide-rest tool. This done, screw your eccentric chuck *one turn* downwards, and turn the screw of your rest *one turn* forwards, moving 12 numbers on the eccentric chuck wheel for every step. As this requires great nicety, delicacy, and care, (the cuts being very deep,) you must no longer depend upon the lever to approach the tool (which is No. 1, of the slide-rest tools) to the work; instead of using it, screw one of the screws, L, at the end of the tool slider, to the deepest cut, and leave it so; then, as you push the tool against the wood, gently unscrew the other screw, L, till the first one prevents its cutting any deeper. For every step, move exactly the breadth of the tool. If the piece you are working is thin and long, it will be necessary to use the back puppet; in which case remember to draw it away from the wood at every change of movement; for, of course, when the eccentric chuck is altered, the centre of the work becomes altered also. Nos. 1 and 2 of plate 13 are different patterns for pillars, both worked *without* the eccentric chuck. The depth of the cuts, it will immediately be seen, form the steps. The rings now demand our attention; the tool used for forming them is No. 6 of the slide-rest tools, plate 5; one side of the tool cuts the outer; the other, the inner part of the rings. Put the tool into a handle, where it must be held firm by a screw; the same handle will be very useful to hold the slide-rest tools for making mouldings and delicate ornaments. The end of the ring tool, No. 6, being very sharp, cuts a road for itself into the wood or ivory, then press it gently to the left, till you find one half of the ring is formed, which will be seen by its filling up the half circle of the tool; then remove the tool to the outside of the work, and press the other side against it till the ring falls off. It is a good way to turn a small cup and line it with velvet, or some soft article, into which to allow the rings to fall, as the delicate ones would be apt to split by striking against the steel bed of the lathe. To form the chain, cut the rings carefully through on one side with a sharp penknife, and slip them one through the other. The handles of the temple are large rings, the pins they hang upon are of wood. Flatten the knob either with a file or a small saw, and drill a hole through it to receive the ring; the pin is then glued into the moulding.

The ball at the top is turned quite round, and may be ornamented with the circular rest. The ball pendant from the chain is circular, except on one side, where a small pin is left, through which drill a hole and slip the ring through it. All the ornaments, pillars, and flutings may be made of separate small pieces of ivory, which, when worked, are easily fastened together by a strong cement made of isinglass melted in gin. The top of the temple may be worked with the universal cutter, or the drill, and the feet are turned quite round, and then ornamented with the eccentric cutter, as directed for the stopper of the bottle, B, plate 9. The needle-case, E, in plate 13, is also a specimen of curious turning; the whole thing should be of ivory, or else the twists of ivory, and the pillar in the middle of cocoa, or some dark wood. Turn the ivory perfectly cylindrical, and hollow it very carefully; this done, throw your eccentric chuck four times outwards, the screw of the sliding rest one turn downwards, and the eccentric chuck wheel four numbers for each plate. The other twists are each worked exactly the same, and they must have a small pin left at either end, which is glued into the moulding. When your centre pillar is hollowed, fit the twists to it.

Another method for working this pattern is with the aid of the drill, and by this means the twists are in the *same* piece of ivory as the pillar. Having formed a smooth cylinder, hollowed it out, and made the screws, take your largest flat-ended drill, and drill fifteen holes round the pillar, or one at every eighth number of the eccentric chuck wheel. Your ivory must be very thick, to enable the tool to cut very deep. Having cut one row of holes, slowly and carefully, you must move the sliding rest forward two turns, then cut another hole every eighth number again; do a third circle the same, moving the chuck and slide the same every time. Now examine your work, and see whether the twists begin to appear. They should, as in the plate, be quite *separate* from the ivory tube in the middle, being only joined at the top and bottom. For this no exact rule can be given, as it depends on the circumference of the work: if very large, the tool must cut very deep to detach it: but, after trying two or three times by these directions, it will be found quite easy. As the ivory twists are, of course, very delicate, and the least jar in working is apt to break, do not use the lever, but employ the screw guides, as directed for the spiral turning. *Line* the ivory before beginning the twists, and leave about half an inch at each end, which you can ornament with the cutter: this gives strength to the work. For the same reason, it is advisable to cut one piece of your ivory longer than the other: on the shorter one make the outside screw, and below it about half an inch of ornament; on the other, make the inside screw, and work it with the cutter to correspond with the other part, quite beyond where the screw extends. When screwed together, it will not be perceived that the two pieces are not even; and as the part where the inner screw is worked is, of course, thin, you would have no thickness for the spiral pattern. When one end is finished, put a fine double angular tool into your cutter, and very carefully cut some circles at regular distances between the twists: this adds much to the curious effect of the patterns: to do it, work backwards, counting the same numbers on the chuck wheel and on the sliding rest as for the twists.

Patterns that are not cut deep can have the impression taken off on paper, either by slightly rubbing the wood over with printers' ink, and laying upon it silver paper, which must be pressed down upon it; or by laying the paper on the wood, and rubbing it well with a piece of lead melted into the shape of a pencil. Patterns on ivory also look very well by rubbing a little printers' ink well over them till the lines are full of it, then wipe off the superfluous ink from the surface, and the design will be black on a white ground.

An immense variety of most beautiful and intricate patterns are worked with the eccentric cutter, in conjunction with the eccentric chuck; indeed, many of those formerly supposed to be only done with the double eccentric chuck, invented by Mr. Ibbetson, can be worked by these two when used together, but they require much patience, and knowledge of the powers of the eccentric chuck, to enable the learner to use them properly. When the eccentric cutter is used, always set the tool exactly in the *centre* of the work, at its lowest degree, before you begin any pattern, or you will never be quite sure that your work is done straight; when once you have settled the centre, the patterns will diverge regularly and evenly from it. In the patterns we are going to describe, the slide rest fixes the position of the tool; the chuck must be screwed down to the proper situation to meet it, and the cutter tool must then be screwed out to cut the circle required.

PATTERN 1, <u>PLATE 14.</u>

This pattern is formed of four groups of circles, containing seven in each. Your wood being perfectly smoothed and *lined*, set the cutter tool (a double angular one) at its lowest degree, exactly in the centre of the work, so that if moved it would only cut a dot. Unscrew it eight or ten turns to make a large circle, then by impelling the cutter box forwards with the slide-rest screw, fix it so that the tool cuts the circle a little *over* the centre: arrange your screw guides: having cut one circle, count 30 on the eccentric chuck wheel, and the same for the other two. It will be observed that these four groups go in straight lines, each towards the edge of the work; to perform this, in the six following rows of circles, move the slide rest forwards half a turn, and the small wheel of the eccentric chuck half a turn upwards for each row of four circles, counting 30, as before, till the twenty-eight are all done.

PATTERN 2.

Plate 14.

Arrange the circle a little smaller than for pattern 1, and on a level with the centre point of the work. Set the chuck wheel at 120, cut a circle; then one at 15, 30, 45, 60, 75, 90, 105, when you will find the eight *outer* circles of the pattern are formed. Then move the slide-rest screw forwards four numbers, and to No. 1 on the eccentric chuck wheel; then to Nos. 14-31, 44-61, 74-91, 104. By following the same rule in the other circles you will find the pattern is not difficult, though at first it may appear very intricate. As all depends upon properly counting the numbers on the eccentric chuck wheel, to make the calculation more easy, I give the proper numbers for the sixteen circles composing one cluster, and you will then see that the others can readily-be done the same, always remembering to move the slide-rest screw forwards four numbers for every row; 1, 15-2, 14-3, 13-4, 12-5, 11-6, 10-7, 9-8; the latter single number forms the centre or outer circle near the edge.

PATTERN 3.

This pattern requires great attention and care to work it properly. Having arranged the cutter tool, screw it out four turns to make the circle the proper size, then move the cutter to the edge of the work, cut a circle at Nos. 120, 40, and 80, on the eccentric chuck wheel, then set the wheel at 20, and leave it fixed. Your next operation must be to screw down the eccentric chuck, and alter the position of the tool on the slide-rest, till, by laying your hand on the fly-wheel and moving it gently, the tool of the cutter appears to describe a half circle *across* the wood, in the same way as for the arc patterns. Having arranged the first arc, move 40, as before, and see whether the second will exactly meet it, then the third; they should each diverge from the centre of the first circles which were cut round the edge; if they do not exactly fit, move the chuck up and down, and the slide-rest screw backwards or forwards, till you find the exact position; as, however, the arc is difficult to settle, the line being of course an imaginary one, as the tool must not touch the wood, it is a good way to cut a piece of pencil to fit like the tool into a box, and with it mark the arcs; they will easily rub out, and thus you will be able to be more certain of your proper distance and position. The next thing is, to see on the fly-wheel of the lathe how many numbers are required to form the arc; thus, if it begins, as the one I have worked did, at 360 of the brass wheel, and ended at 85, stop the wheel at the former number, cut a circle, then one at every fifth number on the same wheel, till you arrive at 85; then move forty numbers on the eccentric chuck wheel, and do the other arc the same; then the third one.

PATTERN 4.

The arc pattern is first worked: arrange it as directed for pattern 3, and mark the outline with a pencil: in this, the arc began at 280 of the fly-wheel, and ended at 168, making in all 116 numbers; to divide 58 for each half of the arc, divide them thus—4, 5, 6, 7, 8, 9, 10, 3, 4; then count back 4, 3, 10, 9, 8, 7, 6, 5, 4. You will see that the numbers increase one every time or for every circle, which also is enlarged one notch for each. Having made the three arcs fit by marking their position with the pencil in the cutter box, cut a very small circle at 280, enlarge one notch, count 5, cut another, enlarge a notch, count 6, and so on till you have counted, in all, 49 numbers; do not enlarge the circle, but cut one, counting three numbers, then 4, which will be the middle, then 4 again, then 3, then 10; after 10, decrease the circle one notch for every cut, and count backwards, 9, 8, 7, 6, 5, 4. The other three arcs are done the same, by counting 40 for each on the eccentric chuck wheel; for each of the crowns cut five circles just above the arc, by counting one number for each circle on the eccentric chuck wheel, then lower the chuck one number, and cut the same circles again through the others; lower it half a turn more, and cut two circles; then half a turn again, and cut one circle above and between the last two.

PATTERN 5.

Set the cutter to the middle, enlarge the circle four turns; then turn the slide-rest screw two numbers to the right, cut one circle, and for each of the other eleven, screw the chuck down one turn, then return to the middle and cut eleven the other way. When 23 are finished, screw the slide rest to the left four numbers, and cut 23 more exactly in the same manner. To work the side patterns begin by that on the right, set the cutter to the middle of the work, unscrew it two turns, move the cutter slide twelve turns to the right on the slide-rest, cut the middle circle, enlarge one number of the cutter, lower the eccentric chuck one turn, and, to keep the *outside* of the circles in a straight line, move the screw of the slide-rest inwards half a number: proceed the same till the nine circles are cut, then return to the middle circle by counting back nine turns of the eccentric chuck, nine numbers of the cutter, and four and a half numbers on the slide-rest; proceed exactly in the same way to cut the other eight circles, only *raising* instead of *lowering* the eccentric chuck one turn for each circle. The opposite pattern is worked the same, only taking care to turn the slide-rest screw outwards the half turn for every circle, or the straight pattern will come contrary.

In all these straight patterns take great care that the fly-wheel is stopped in such a manner that the eccentric chuck stands perfectly upright: to do this, hold the T square against one side, and make a mark on the brass wheel for the stop to enter. When you are quite sure that the chuck inclines neither to the right nor the left, drill a small hole in the wheel sufficiently deep for the stop to hold it quite firm. You will then find your patterns will always come straight.

PATTERN 6.

Set the cutter to the centre, enlarge the circle six turns, cut one, diminish one number on the cutter for each of the other eight circles, and move the eccentric chuck downwards one turn for each. Count back to the middle circle, and work the other eight the same. The side patterns are done much like those in pattern 5, only the large circle is in the middle, and the straight lines incline inwards.

PATTERN 7.

Set the cutter to the middle of the circle, enlarge it two turns and a half, cut a circle in the middle, turn the chuck down two turns, so as to cut another a little into the former ones; continue the same till the seven are done; return to the middle circle, and cut the other six in the same line, by screwing the eccentric chuck upwards instead of downwards. When the thirteen are completed, lower the chuck one turn, and screw the slide rest outwards (if doing the lines to the right) two turns, cut a circle, then move the eccentric chuck downwards as before, till the second line of circles is cut, each line diminishing one turn of the eccentric chuck at the top and bottom.

PATTERN 8.

Set the cutter to the middle, enlarge it two turns, turn the slide rest outwards twelve turns, cut a circle near the edge; count 20 on the eccentric chuck wheel, cut another, then 5 more, each distant twenty numbers from each other. To make the circles join in straight lines, screw the eccentric chuck downwards two turns for every one; when the outside pattern is done, return to the middle circle, and cut the other lines the same.

PATTERN 9.

Square patterns require great care in working them. First saw the wood perfectly square; then, when on the lathe, take the T square; hold the flat edge firm on the bed of the lathe, and the handle against one side of the wood, till both are even, then fix the wheel of the eccentric chuck to that number. Having with the cutter worked one row of patterns, move the chuck 30 numbers, which, if the wood is quite square, will enable you to do the second row; then 30 more for the third side, and 30 for the last.

We are told, that French and Italian turners often line boxes with the peel of the Bergamot orange; they cut a circle through the peel, and carefully strip it off in two quarters, turning the inside out, and drying them; the scent is very powerful; of course they can only line globular boxes the size of themselves.

TO LINE WOODEN BOXES WITH TORTOISE-SHELL.

As many of my readers may wish to line their snuff-boxes with tortoise-shell, I think it needful to give some instructions in the best method of doing it. Cut the shell into very thin leaves with a sharp fine saw, then divide these into the size you wish for the inside of the snuff-box, leaving a very little extra for the joint. Take a new rough file, and scrape away a little of the two *ends* of the narrow strip of shell, so that they can lie one upon another, and fit so closely that the aperture is hardly perceptible. Plunge the tortoise-shell into warm water for a few minutes, and it will become quite soft. Have ready on the lathe a piece of wood a little *less* in circumference than the inside of the box, and perfectly round and smooth. While the shell is soft, place the joints together, wrap a wet piece of linen tightly over them to hold them fast, and press them firmly together with the finger and thumb. Then heat a pair of tongs to a proper heat, (which is known by trying them on writing paper; if they brown it, they are too hot; if they only turn it yellow, they are right,) and with them compress the joint of the tortoise-shell. The water, the heat, and the pressure united, will make the two parts join firmly. When finished thus far, file away any roughness that may remain, and steep the shell into hot water till quite soft, then slip it upon the piece of wood before mentioned, and see if it fits it perfectly; if not, try with pressure to give it the requisite shape; and if this does not succeed, take the wood out of the lathe, leaving the tortoise-shell upon it, and hold them over a brazier, turning them frequently and quickly between the hands, that the heat may equally penetrate all the parts; then strike the side that bulged out with a mallet, and with a little care it will soon assume the required form. You may finish it on the lathe, observing only to place it so that the tool does not catch the lap of the joint, which might cause it to open; and when you take it off, plunge it in cold water to make it retain its form.

TO FINISH THE SNUFF-BOX.

Make your box and lid of hard, well-seasoned wood; hollow them out, and polish the insides, only omitting to cut the lip upon which the lid fits, and which will be formed by the tortoise-shell. As you hollow out the box and lid, keep fitting in the shell, that you may not make them too large; when it slips in rather tight, take a point tool and cut some circles on the inside of each; (this is done to enable the glue to hold firmly.) Now, take the tortoise-shell and file that part that is to be cemented to the box, so as to make it rough. Take a pair of compasses, in one end of which is a sharp knife, set them to the exact size of the inside of the bottom and lid, then place them on a piece of tortoise-shell, and cut out the two round pieces. Melt some glue till rather liquid, thicken it with vermilion, lay a coat on the inside of the top and bottom of the box, and press in the two circles of tortoise-shell; in the same manner glue in the sides, leaving the lip (upon which the lid is to fit) standing up above the bottom part of the box. Leaving the glue to harden for a day, then replace the work on the lathe, turn the inside quite even, and polish it with pounce powder and oil, then with tripoli powder and water. Should the lid, when finished, become too small for the box, dip it for a minute in boiling water, fit it on to a piece of wood the exact size, and leave it there to harden.

HORN TO IMITATE TORTOISE-SHELL.

Dissolve three ounces of potash in a pint of boiling water. Let it boil for a quarter of an hour, then pour it into a basin capable of holding about as much again, and in which you have put half a pound of quick lime, stir it well, and when the latter is *slacked*, add three ounces of red lead and one ounce of vermilion.

When the whole is of the consistency of thick soup, dip a thin pointed stick into it, and lay the drops it will take up upon a piece of horn in those parts required to be coloured, leaving those that are to be transparent. When quite dry, clean the whole with a wet sponge, and you will find it will greatly resemble tortoise-shell.

MASTIC USED IN TURNING IVORY VERY THIN.

To turn ivory as thin as writing paper, so as to render it quite transparent, is very difficult to accomplish, but is much admired when done, and shows the skill of the artist. To enable the ivory to bear the action of the tool without splitting, the following mastic has been found very useful, both for strengthening it and for giving a deep colouring, by which means the thinness is more perceptible. Some turners wet the ivory for the latter purpose, but as when wet it is quite transparent, and thickens again when dry, the mastic will be found much preferable.

Take some lamp or ivory black in powder, and strain it through a fine sieve, so as only to retain the finest parts. Steep these in water to free them from any impurities. After lying in it a few minutes, pour off the water, and make some glue very hot, mix it with the lamp black till of the consistency of oil paint. This mastic must be kept warm near the fire, and when you have sufficiently hollowed out the vase, or whatever you wish to turn, very thin, shape the outside a little: then dip a large camel's-hair brush in the warm varnish and lay a thick coat all over the inside; let this dry, put on another, and repeat the process till sufficient strength is obtained. You may now, without fear or danger, work your ivory as thin as possible, and ornament it with the cutters and drill. Without this mastic it would not, when transparent, bear the force of these tools.

When the work is all finished and carefully polished, take it off the lathe, and put it in a basin filled with warm water. After a few minutes' immersion, take it out, and plunge it in clean water, shaking it gently. This will make the mastic dissolve and leave the ivory. Renew the warm water frequently, as leaving it in the blackened liquid might injure the colour of the work.

BEAUTIFUL VARNISH FOR WOOD, TO BE USED WHILE THE WORK IS ON THE LATHE.

To one quart of spirits of wine add four ounces of lacker, three ounces of gum benzoin, one drachm of camphor, half a drachm of sandarac, half a drachm of dragon's blood, one drachm of turpentine. Put these ingredients in a long-necked bottle capable of holding two quarts, and tie a piece of wet parchment over the neck; when dry, pierce it with holes with a large pin. Place the bottle in the *bain marée* till the contents are perfectly dissolved, shaking it frequently. When cold, strain the liquid through a piece of coarse muslin, and keep it well corked for future use.

Having finished and polished your work with tripoli powder and sand-paper, wipe the wood quite clean with a piece of fine linen. This done, put a few drops of the varnish on a bit of cotton wool, and one drop of olive oil, to prevent its drying too quickly; while applying this varnish, make the lathe wheel go very quick, and hold the cotton close to the work.

To polish the above, when the varnish is perfectly dry, take some finely-powdered whiting or chalk, and with it polish the work in every direction, but be careful not to press too strongly on the varnish, or it will be marked; and do not rub it for more than a few minutes at a time, as the friction and excessive speed and heat will spoil the polish. When done, take a sponge dipped in water, wash the work well, and then rub it with a piece of fine linen, and a drop or two of olive oil; lastly, clean it with a bit of soft old rag or leather.

TURNER'S CEMENT.

Sometimes the workman is too much hurried to wait till the work can be glued upon the chuck; the following cement will be found useful. Take two pounds of Burgundy pitch, one pound of rosin, one pound of colophonium, two ounces of yellow wax, and one ball of whiting. Melt all, except the latter, in an earthen pot, over a slow fire. When it begins to bubble, stir it well with a stick to prevent its passing the edge of the pot, and when all is quite melted take it off the fire. Add the whiting, finely powdered, little by little, stirring in well till the contents are perfectly mixed. Replace the pot on the fire, still stirring it; and after a few minutes, pour all the cement quickly into a tub or basin of cold water for about a minute. Then take it out, and knead it well with the hands. Roll it into sticks upon a smooth stone, and plunge them into cold water to harden. The strength and goodness of this cement depends greatly upon its being made as rapidly as possible. To use it, melt the end of one of the sticks by putting it near the fire; rub it on the chuck; when you think there is sufficient cement laid upon it, heat the bit of wood or ivory you wish to unite to it, and the warmth will make it adhere firmly.

Another cement, that is preferable for using in cold weather, is made by adding two pounds of Burgundy pitch, two ounces of yellow wax, and two pounds of Spanish white. These are melted together, rolled into sticks, and used as above; one stroke of the mallet will detach the work from the chuck when joined with this cement.

NEW AND VALUABLE RECIPE FOR TAKING BEAUTIFUL IMPRESSIONS FROM TURNING PATTERNS.

Take a sheet of rice-paper, paste it upon letter-paper with flour paste, which must be mixed as smooth as possible, and laid on very thin. Leave it till quite dry, then lay the rice-paper thus backed upon the piece of Turning, and with the thumb nail or a piece of cloth rub the back of the paper, pressing it gently so as to make it enter into all the deep, fine cuts of the Turning. The impression will be beautiful, and have the appearance of a raised medallion. It is invaluable for taking off patterns from ivory, as it can in no way injure the colour or delicacy of the work; and from the facility of bending the paper, impressions can as easily be taken from round articles (as the sides of a box or pillar) as from flat ones. The medallions may be left white, or the pattern coloured with water colours, leaving the ground white; and they can, besides their utility as patterns to which to refer, be employed in ornamenting various useful articles.

In turning the pillars in the temple, pl. 13, great inconvenience and trouble have been found in working them, for want of the support of the puppet G, plate I; for as in each step the eccentric chuck has to be slightly altered, the point, J, of the puppet, when the work is moved, is apt to slide back into the hole it first formed, thus making the steps quite crooked. To obviate this difficulty, I should advise my readers to use the following simple contrivance. Take out the point, J, and replace it by a piece of wood turned to fit exactly into the tube, and on the end of which is left a circular piece the size of a half-crown. Take another bit of wood, of the diameter of your pillar; fix a short nail into the middle of it; glue the other side to the pillar, put it on the lathe, screw up the puppet, G, and you will find the nail will always, whenever you move the chuck, firmly fix itself into the wooden end which replaces the point J.

And now, having, I trust with sufficient clearness, explained the practical and ornamental parts of concentric and eccentric Turning, I will take leave of my readers, only adding that, if they wish to attain perfection in this interesting art, they must patiently continue their exertions, for experience and industry will alone enable them to avoid many faults, and discover the real cause of many failures; and to those who may feel disheartened with repeated disappointments, I will say—

"Courage! try thy chance once more."

Let me also observe to those who cannot afford to purchase much expensive machinery, that with care, patience, and perseverance, the common tools may be made to work a great variety of very beautiful articles: and if they will also keep in mind the old-fashioned but true saying, "that whatever is worth learning at all, is worth learning well;" they will, I have no doubt, soon become proficients in an art that has been admired and practised for centuries.

THE END.

Milton Keynes UK
Ingram Content Group UK Ltd.
UKHW030839021124
450589UK00006B/676

9 789362 515063